Hands of Time

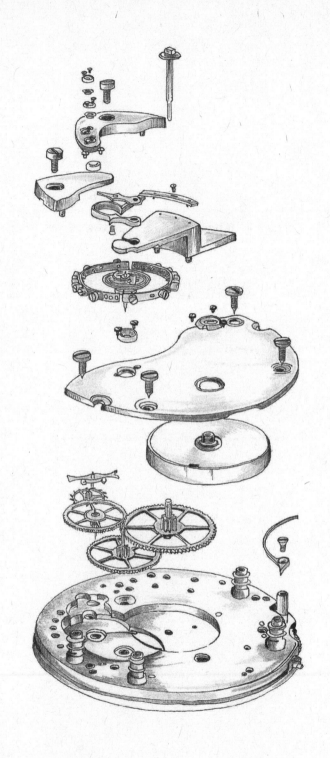

Hands of Time

A Watchmaker's History

REBECCA STRUTHERS

WITH ILLUSTRATIONS BY CRAIG STRUTHERS
AND PHOTOGRAPHS BY ANDY PILSBURY

HARPER

An Imprint of HarperCollins*Publishers*

HarperCollins books may be purchased for educational, business, or sales promotional use. For information, please email the Special Markets Department at SPsales@harpercollins.com.

Originally published in Great Britain in 2023 by Hodder & Stoughton, an imprint of Hachette UK.

Extract on p. 145 from "Attack" by Siegfried Sassoon, copyright © Siegfried Sassoon and reproduced by kind permission of the Estate of George Sassoon.

Illustrations © Craig Struthers 2023
Photographs © Andy Pilsbury 2023

FIRST U.S. EDITION

Library of Congress Cataloging-in-Publication Data has been applied for.

ISBN 978-0-06-304870-6

23 24 25 26 27 LBC 5 4 3 2 1

In memory of Adam Phillips and Indy Struthers

Contents

A Backward-facing Foreword   ix

1   Facing the Sun   I
2   Ingenious Devices   17
3   Tempus Fugit   36
4   The Golden Age   53
5   Forging Time   76
6   Revolution Time   93
7   Working to the Clock   III
8   The Watch of Action   130
9   Accelerated Time   145
10   Man and Machine   165
11   Eleventh Hour   188

How to Repair a Watch   197

Glossary   207
Acknowledgements   215
Picture credits   219
Bibliography   221
Further Resources   233
Notes   239
Index   253

A Backward-facing Foreword*

Jeff 10/3/71

Whoen I first started training as a watchmaker at nineteen, I was taught never to leave a trace of my presence within the watches I restored. And yet, these traces can tell us all kinds of stories about otherwise inanimate objects. For example, the vintage Omega Seamaster wristwatch on my workbench was repaired by a man called Jeff on 10 March 1971. I know this because Jeff scratched his name and the service date into the back of the dial, so that if the watch ever came back to him, he would know that he had previously worked on it and when.

Working as we do on objects only a few centimetres in diameter, a watchmaker's world is often not much bigger than a thumbnail. It is all-consuming. Sometimes a whole morning passes and I have barely shifted my gaze beyond the postage-stamp-sized mechanism I am working on. I suddenly realise the coffee next to me is cold and my eyes are dry from concentrating so hard I've forgotten to blink. My husband, Craig, is also a watchmaker and, although we work on benches that face each other, we can spend whole days in near silence, exchanging little conversation beyond orders for the kettle. When we make a new watch, whether from salvaged parts or from scratch, it can take us anywhere from six months to six years. We can measure sections of our lives by these watches, and sometimes find ourselves noticeably older when we are finished with them.

★ With thanks to Alexander Marshack's *The Roots of Civilization*.

Our workshop is in an eighteenth-century goldsmiths' factory in Birmingham's historic Jewellery Quarter. Craftspeople have been producing work here for seven generations and the walls feel steeped in appropriate history. The rooms beneath us contain centuries-old presses, dies and design charts, as well as artisans still using this equipment to produce jewellery. Our small room on the top floor is bright and airy, with skylights and arched windows. When we first moved in and were preparing the space, we were told that during the Blitz a bomb had crashed through the roof and failed to detonate. I pulled down an old insulation screen covering the skylight and found a beam still charred from the blast, which I scrubbed and left on show. It now presides over our lathe bench and wheel-cutting machine, which, ironically, is German. We call her Helga. She sits on a long bench that extends down the whole side of our workshop and is covered with a variety of old machines. Underneath, there are copious drawers full of the dull gleam of old watch mechanisms and parts that we have sourced or rescued over the years – often from bullion traders who had taken them out of their cases, which were to be sold as scrap gold or silver, or from the workshops of watch-makers past being cleared out by their families. Our 'clean' watch-making benches are on the opposite side of the workshop, as far as possible from the swarf (metal shavings) and oil that occasionally spit from the machines.

We keep our workshop very clean, to avoid dust or dirt finding its way into the delicate watch mechanisms. In the state-of-the-art watch manufactories of Switzerland or East Asia, workshops will have double air-lock doors and sticky mats to remove dirt from shoes, and watchmakers wear compulsory lab coats and shower caps over their feet. We are a little more relaxed. Our dog, Archie, snoozes in a corner. By the end of a day making new watch parts the room smells of lathe oil, a distinctive aroma almost like that of a tomato vine, with notes of metallic copper and iron. There will be small mounds of brass or steel swarf scattered around our lathes, mills and drills, as well as oil- and coffee-ring-stained sketches of parts strewn liberally across the workbenches. We sweep the floor regularly, ready for the occasional team hunt for accidentally flicked parts – you could probably make up a complete watch from the parts stuck between floorboards or rolling under sets of drawers in most watchmaking workshops. Our floor is pale grey vinyl, the perfect contrast for the yellow of brass or bright red of a ruby jewel.* No one tells you that one of the key skills for any watchmaker is the ability to find tiny shiny things on the floor.†

The last residents of our workshop were enamellers, and traditional makers have worked in this space for over two centuries. In this room at least, not much has changed. Although we have modern computers, most of the tools and machines we work with are between 50 and 150 years old. Our skills too, are from a bygone era. In the 'golden age' of watchmaking that ran through the seventeenth and eighteenth centuries, Britain was the centre of the watchmaking world. Now watchmakers like Craig and me are a rare breed. In 2012 we set up on our own, becoming just one of a handful of firms in the UK with the skills to make mechanical watches from scratch and to restore antique watches from the last five centuries. But the course we trained on no longer exists. The Heritage Crafts Red List of Endangered Crafts (much like the Red List of Threatened Species but for craft) currently lists artisanal watchmaking as a critically endangered skill in the UK.

In part our skillset is disappearing because, in our technologically advanced age, computer numerical control (CNC) can virtually make a whole watch for you. And you might well ask why we bother with this old equipment when we could feed a computer design into a software-operated machine to do most of the manufacturing for us instead. But where's the fun in that? We love getting our hands dirty

---

* Many mechanical watches use bearings made from synthetic ruby, or corundum, as it makes for an incredibly hard surface. The small steel pivots that support each toothed wheel can rotate against the ruby without it wearing away from the friction.

† Our hunts are usually conducted under the watchful, if a little confused, eye of Archie, who has never quite come to terms with the notion of looking for things that aren't edible.

making things and fiddling with little parts to get them to work together. You build a closer relationship with what you're making when you work by hand. You can hear when the cutting speed of a lathe or drill is perfect, and you can feel from the resistance whether the pressure of your tool is correct. We like this sense of connection to the objects, and to the generations of artisans who came before us.

I have always been fascinated by time, but I never set out to be a watchmaker. At school I wanted to be a pathologist (this was long before TV crime dramas made forensics cool). I was an oddball who was fascinated by how things worked, particularly bodies. I wanted to help people, but I wasn't always very good at actually talking to them; working with the dead, I reasoned, would save me a lot of difficult conversations with patients. I liked the idea of figuring out why a body had stopped functioning. In the process I hoped I might help other people, perhaps by helping to bring about justice, or a deeper understanding of a deadly disease.

My career as a pathologist was never to be, but there is something forensic about working on old watches. Watch mechanisms contain tens if not hundreds, and occasionally thousands, of components, each of which has a specific task to perform. The most basic simply tell the time. The most complicated (the added functions in a watch that go beyond time-telling are in fact called 'complications') can chime the hours and minutes on gongs made from finely tuned wire, accurately maintain the date for over a century, or chart the stars. When any of these parts become faulty or need cleaning, or re-oiling, they stop the mechanism from functioning. As restorers we dissect to determine the cause of death, only with the bonus that, once repaired and reassembled, our subject has another chance at life. The final stage of reassembling a mechanical watch is to replace the balance, which makes the watch start to tick again. There is nothing quite like hearing life restored to a piece that has not worked for years, or even centuries, knowing that its tick sounds the same to me now as it did to the watchmaker who first assembled it. The pulse of the balance is referred to as its 'beat', and the coiled spring used to regulate its action 'breathes'.

As time went on, it felt very natural to me to begin switching between working on watches and thinking and writing about them and their history. I became the first practising watchmaker in the UK to pursue a doctorate in antiquarian horology (the study of the history of timekeeping). After all, restorers are, in part, historians. It's a practical kind of history: you have to know how something was made and how it once operated to return it to the way its maker intended. Now I found it worked the other way too: when Craig and I first started making our own watches from scratch, my historical research and writing influenced the watches we made, in a sort of horological cross-fertilisation. My research enlarged my tiny watchmaker's world. The focus of the watchmaker is often smaller than a grain of rice, but the inspiration for horology is the universe – I love this contrast of micro and macro. And poring over the construction of an eighteenth-century watch to discern what it could tell me about its provenance and owners made me keenly aware not only of how history had shaped the watch, but also of how the watch has shaped us.

It would not be a stretch to say that the invention of mechanical timekeepers has been as significant for human culture as the printing press. Imagine trying to catch a train by relying on the position of the sun. Or organising a Zoom conference of 200 people located all around the world, each trying to decipher the start time by hanging out of their window to be within earshot of the bells of the nearest public clock. Or, at the more life-and-death end of the scale, think of surgeons performing an organ transplant or removing a tumour with no accurate reference point to measure their patient's heart rate. Our ability to do business, structure our day, and access life-saving developments in science and medicine all rely on – are, in fact, made possible by – access to accurate time.

From its very beginning the watch both reflected and developed our relationship with time. Watches don't create time, they measure our cultural perception of time. All time-measuring devices, whether they are ancient carved bones or the watches I restore on my workbench, are a way of counting, measuring, analysing the world around us. The earliest timekeepers began by tracking naturally occurring phenomena in the world and solar system. Even now, the most up-to-date modelling devices we possess – smartwatches such as the Apple Watch – can still track a celestial routine, keeping pace with our planet as it hurtles round the sun over the course of a day. The systems we've developed to understand these processes, and our place within them, are our way of getting to grips with our universe, of applying a cosmic rational order that we can use to better live our lives.

What we call a watch – a tiny, wearable clock – is a miracle of engineering. Mechanical watches are among the most efficient machines ever created. I have worked on watches that haven't been serviced since the 1980s, and yet have only just stopped running. I struggle to name another mechanism that would work day and night for nearly forty years before requiring any maintenance from a mechanic. As of 2020, the most complicated watch in the world contains nearly 3,000 parts and is capable of measuring the Gregorian, Hebrew, astronomical and lunar calendars, chiming the hours and minutes together with fifty other complications, all in a device that fits in the palm of your hand. The smallest watch movement ever created was first made in the 1920s and fits ninety-eight parts into a volume of only 0.2 cubic centimetres. The first chronometer, a watch so accurate it could be used by sailors to calculate longitude at sea, was made over sixty years before the invention of the electric motor and over a hundred years before the first electric lighting. Watches have since accompanied humans to the summit of Everest, the depths of the Mariana Trench, both the North and the South Poles, and even to the moon.

Our concept of time is inseparable from our culture. In fact, the word *time* is the most commonly used noun in the English language. In Western, capitalist cultures time is something we have, or don't have, save or lose, it marches on, it drags, seems to stand still and flies. Time thrums constantly underneath everything we do. It is the backdrop and the context for our existence and our place in what is now a supremely mechanised world.

Slowly, over the course of tens of thousands of years, the power balance between humans and time has slowly shifted. What began with us living our lives around the natural phenomena the world threw at us developed to become an entity we have sought to control. Now it often feels as if time controls us. Time, we've discovered, is not as 'fixed' as we first believed. It might not be universal, constantly moving on, waiting for no man. It might be relative, personal and even, one day, reversible – medically speaking at least.

I knew early on that I wanted to be a watchmaker rather than a clockmaker. Watches have followed us through our daily lives for centuries, worn on or near our bodies. I've always been fascinated by the intimacy of that relationship. The connection between person and watch, its tick reflecting the beat of our very own pulse, the rhythm of our own body, was for a long time the closest relationship we had with a machine – until, of course, mobile phones came along. In many ways watches are an extension of us, a projection of our identity, our personality, our aspirations, as well as our social and economic status. A watch is an individual's timekeeper, but it is also a kind of diary: it holds in its restless hands our memories of the hours, days and years we have spent wearing it. It is an inanimate but uniquely human repository of life itself.

This book, too, is a history of timekeeping, and of time itself, but from my irregular perspective as a twenty-first-century watchmaker. We begin with the very earliest human-made timekeepers, crafted from bones, measuring shadows, channelling water, fire or sand. I'll then explore how inventors later found ways to combine natural power sources with artificial engineering. The first clocks as we understand them were a product of an extraordinary combination of curiosity, experimentation and highly sophisticated science. Their mechanisms, which were once so huge that they could only be accommodated in massive church towers, were the forerunners of the miniature timekeepers I handle every day on my workbench.

From there we'll turn to the wonder of watches. Each chapter explores a pivotal moment in the history of the watch, from its advent 500 years ago to the present day. I'll unpick the mind-boggling technical advances that enabled these machines to become portable and accurate enough to conquer the world. I'll show how these little instruments coordinated work, worship and wars, and how they helped European nations navigate and map the world, supporting global trade and enabling colonial expansion. We'll show how explorers and battle-weary soldiers depended on them for their very survival, and how decisive historical events were dictated by them. We'll also track their evolution from elite status symbol to popular tool to status symbol all over again. The watch is the metronome of Western civilisation itself, establishing a rhythm that has driven our history and continues to govern our time- and productivity-obsessed age.

It is also a personal history. My particular interests glimmer behind the dial. Many of the watches in this book I have handled, or even repaired, and the stories they tell are central to this narrative. I once worked on a watch, destined to be a wedding gift from parent to child, that had been in the same family since the eighteenth century. Holding it, working on it and considering its past and future, I felt like I was bridging time itself. Handling hundreds of years of horology creates an eerie sense of self-awareness. Working on an antique watch in minute detail, I feel an almost tangible connection to the people who made and wore it. Tiny traces of humanity stand out like signatures – initials or names like Jeff's, concealed within the mechanics, or a 250-year-old enameller's fingerprint, accidentally baked into the blue-green glass and hidden underneath the dial of a pocket watch. Conscious that I am another chapter in the story of an object created before I was born and which, if cared for, will live on for centuries after I'm gone, I collect these signs of life.

A watchmaker is a custodian, protecting these objects, absorbing their history, as well as preparing them for new connections yet to be created. Occasionally I see a watch again, for servicing or repair, years after I've first worked on it, and it's like reconnecting with an old friend. My memories of the watch's marks and idiosyncrasies come flooding back, and sometimes new ones are created. I've had a recently repaired watch returned with water damage; not long after it had been given as an eighteenth birthday present it was submerged in a swimming pool along with its mildly inebriated new owner on holiday in Mallorca. The mechanism inside has now been restored to its original state, but the slight stain around the three o'clock position on its dial is a permanent reminder of the perils of mixing tequila shots, chlorinated water and vintage precision mechanics.

Every morning, when I sit at my bench and start work, the watch in front of me is a new beginning. Each one has its own history. Engineering perfection aside, each knock and scrape, each hidden mark left by a past repairer, even the way they were designed, and the techniques used to make them, are clues to a story that extends far beyond the tiny object before me.

Hands of Time

I

Facing the Sun

Kia whakatōmuri te haere whakamua
I walk backwards into the future with my eyes fixed on my past
Māori proverb

I've always been fascinated by nature. As a child I liked nothing more than gathering slugs in the garden, and getting covered in mud and slime. Above all, I loved to learn how things worked. One of my earliest memories is my dad first showing me his microscope. I was dazzled to discover another world secretly existing within my own but invisible to my naked eye. I loved it so much my parents bought me a children's version for Christmas, which was portable, so I could use it around the garden. We spent hours studying pond-water samples. I then drew the weird and wonderful creatures I'd see darting and crawling around on the slide.

I grew up in a suburb of Birmingham called Perry Barr. It was a densely populated sprawl of brick, concrete and tarmac, sliced in two by the A34 and its undulating flyovers and underpasses. The closest thing to a local landscape was a fly-tipped rubbish patch of wasteland where my sister and I snuck off to play. We called it the 'background' (it was quite literally the ground at the back of our house).

I don't remember much about the seasons in Perry Barr. Aside from the odd smattering of snow in winter, the spiky grass of the 'background' was reddish-brown all year round. In autumn, leaves gathered in slimy clumps on the pavement, while my parents debated whether it was too early to switch on the heating. At night the streetlights suffused the sky with a milky orange glow that made the stars almost invisible.

I will always be a Birmingham girl. But in my early thirties Craig and I had no choice but to leave. We were economic migrants forced *away* from the city, seeking lower house prices within the range of our freelance incomes. We bought an old weavers' cottage in a small town in the northernmost part of Staffordshire, bordering the Peak District. It was the cheapest place we could find within a 50-mile radius of our workshop.

Neither of us had ever lived so close to the countryside. We spent our first months exploring the fields, woods and moors around our new home while walking our dog. Archie's favourite route took us through a valley I later learned was called Little Switzerland – a fitting choice for two watchmakers and their watchdog. Ever drawn to the relics of an industrial past, we liked to stroll along the repurposed railway line that once connected Cheshire to Uttoxeter, through the woodlands of Dimmingsdale alongside the River Churnet. Archie's nose would be aquiver with the unfamiliar natural smells: badger, deer, weasel, owl and vole.

As the seasons changed, so did the walks. In winter, the low rays of sunlight pierced the skeletons of old oaks and frostbitten hedges. In spring, the woodland shadows filled with bluebells. Autumn brought mists so thick that sometimes we struggled to see more than a few metres ahead. I started to notice how animals were rotated in their fields – which times of year the cows would be out, and when the sheep would be lambing. I learned the hard way that Archie had to be kept away from certain areas during muck-spreading seasons: late winter and spring.

My first autumn in the cottage was spent working on an important watchmaking project, with a deadline set for Christmas. It was a particularly complicated and ambitious project and, as the days rolled on and my progress didn't, I kept telling myself, 'The year isn't coming to a close yet, I still have time.' But increasingly I wished

I'd invested my energy in inventing a time machine rather than machines to tell the time.

One late-autumn afternoon, I looked up and saw a flock of Canada geese flying in a rowdy 'V' formation across the sky. As the weeks went on, these flocks got larger and larger until one day, as I was walking through the woods, the whole sky was filled with beating wings and honking beaks. Archie tilted his head from side to side with a curious expression that I presume meant either 'what's that?' or 'that looks tasty: should I chase it?' I suddenly remembered standing in the 'background' as a child, looking up at a similar flock of geese. For a brief, bitter-sweet moment, past and present collided.

In the northern hemisphere flocking geese are a reliable sign that the year is coming to an end.* With my deadline looming, my strongest feeling was that I wanted them to stop; it was almost as if they were telling me that *my* time was running out. We were both, in a way, keeping time.

Archie watching a flock of Canada geese.

* I had always understood that geese were migrating. In fact, Canada geese are generally resident in the UK but still flock in the autumn.

The natural world surrounding us is riddled with temporal cues if you know where to look for them. It was our first clock, and it continues to tick around us for those that take notice. It was living with and saturated by nature that caused humankind to develop the first timepieces. If watches are personal time, then our first watch was our internal one. You could say that the watch emerged from our first efforts to align our inner sense of time with what we observed in the world around us.

The object that archaeologists currently consider to be the strongest contender for the earliest known timepiece is 44,000 years old. It was discovered in 1940 when a man collecting bat guano in modern-day South Africa discovered a cave in the Lebombo Mountains, nestled among the bush and scrub. The cave was filled with ancient human bones, some of which were 90,000 years old. The site, now called Border Cave, is one of the most important in the history of humankind. Border Cave, which was continuously inhabited by humans for 120,000 years, protected its inhabitants both in life and in death. Located high in the mountains overlooking the plains of what is now called Eswatini, it was an easy site to defend from predatory animals and other humans and offered a good vantage point from which to scout for prey. Archaeologists found more than 69,000 artefacts there, many showing a rich understanding of the natural world and how to engage with it: there were sticks used to dig for carbohydrate-rich tubers, sharpened bones for leatherworking, pieces of jewellery made from ostrich eggs and marine shells, and straw bedding that, repeatedly layered on top of ash and camphor bush, was probably used to repel biting insects and parasites like ticks.

But for me, the most extraordinary discovery was a small piece of carved baboon fibula, about the length of an index finger, inscribed with twenty-nine clear notches, and polished by the hands of its owners through many years of use. It is the first clear evidence of calculation in human history. The Lebombo Bone dates back to long before the advent of agriculture or any evidence of planning for the seasons, and even longer before we conceived of anything close to a regular working day. It is a measuring device from a time when, so far as we know, there was very little to measure.

So what were our ancestors trying to calculate? We can't know for sure, but some scholars have a theory. After the passage of day and night, another likely division of time for our ancestors is thought to have been the phases of the moon. The marks on the bone consist of thirty spaces alternating with twenty-nine notches. The average lunar month is about 29.5 days. If our ancestors rotated measurement between the notches and the spaces that divided them, they would have reached the average of 29.5 days and therefore correctly calculated the lunar month. Some scholars have even suggested that its makers used the bone to track their reproductive cycles, or the length of a pregnancy. I like to think it was handled by our great-great – hundreds of greats – grandmother as she counted down the days.

Many ancient cultures believed that the two cycles – lunar and menstrual – were linked; indeed the belief persists to this day. A recent study found no definitive link, but hypothesised that modern lifestyles, particularly our exposure to artificial light, might have weakened the synchronicity. If so, we would be far from the only creatures whose internal clocks are aligned to the rhythms of the natural world.

My friend Jim,* a farmer and master whisky blender in western Scotland, and his wife Janet, a fourth-generation shepherdess, told me how the sharply decreasing daylight hours in November trigger their ewes to ovulate. With incredible predictability and to within a few days, all the ewes in the flock will follow virtually the same

---

* I got to know Jim through our mutual love of whisky and craft, one of the many unusual and fascinating connections you form over the years when you're in an equally unusual and eclectic line of work.

cycle. Within two cycles, most if not all of the ewes will be pregnant. Twenty-one weeks later, in around the first half of April, the lambs are born, aligning perfectly with the end of the bitter winter weather and the beginning of spring growth. Jim describes this work of coordination as 'getting the mouths ready to eat the grass'.

Just as the lambs arrive, one or two days either side of 17 April, swallows arrive from their 6,000-mile migration, escaping the heat of the South African summer. Between the new livestock and the nesting swallows, spring breaks on the farm in what Jim lovingly refers to as a 'huge burst of life'. In September the swallows queue up on telegraph cables and tree-branches, knowing, somehow, that it is time to leave.

Every living creature has an internal clock. Those of us who live with canine companions and work a regular schedule will have noticed their eerie ability to predict when we're about to come home after a day at work. This is thought to be because the moment we step out through the front door we leave a human smell timer for our dog, who learns that once our scent has diminished to a certain level, enough time has passed for us to be due home. The rooster crowing at daybreak, a timekeeper across the globe, is operating according to an inner circadian clock that has been found to run for an average of 23.8 hours, hence their crowing just before the dawn. Even organisms as small as plankton are known to move up and down in the water, from the depths to the surface, every dusk and dawn. We can be fairly certain that they sense changes in UV levels (in very strong sunlight they sink a little lower into the water to avoid damage) and so can tell day from night by the sun's light. They have even been observed, in controlled experiments in a dark aquarium, to continue to make their vertical

'Oh my, is that the time already?' A plankton heads to the surface.

migration for several days in the complete absence of light. In other words, they too have a biological clock that functions around a 24-hour system.

If anything, our ability to read our internal clock is weaker than most animals' because it's interfered with by our perception of time, which can be warped by emotions: happiness, novelty and absorption seem to speed it up, while boredom and fear appear to slow it down. This body clock undeniably exists, almost like a sixth sense, but it is not universally understood (my internal hour may not be your internal hour). Timekeepers are a symbol of our drive to share, quantify and externalise our intuitive awareness of time. The Lebombo Bone suggests that we were doing this as early as 40,000 years ago.

Ancient measuring devices have been found on almost every continent. Many of the oldest examples sprang up independently of each other, and their patterns suggest different purposes. The first humans to populate Europe, the Aurignacians, left behind what appear to be early calendars. In Baden-Württemberg, a small plaque made from an eagle's wing bone is believed to be the world's oldest known star chart.* In the Democratic Republic of Congo, a 25,000-year-old bone tool handle, known as the Ishango Bone, bears a series of carved notches suggesting mathematical calculations like addition, subtraction, doubling and prime numbers.

These hand-held devices seem to mark an important conceptual turning point for our species. As the philosopher William Irwin Thompson puts it, 'The human being was no longer simply walking in nature; it was miniaturizing the universe and carrying a model of it in its hand in the form of a lunar calendrical tally stick.' But I think they do more than this. By capturing cosmic events in a device that we can put on our wrist or hold in our hand, we are reassuring ourselves – perhaps misguidedly – that we can control the

---

* The eighty-six notches could represent the number of days that one of Orion's two prominent stars, Betelgeuse, is visible.

uncontrollable. They make us feel that we are no longer just existing in time but *using* it, to our advantage.

What did time feel like to ancient humans? Did they simply live 'in the moment', the fantasy of many a self-help devotee? It may well be that they lived in 'survival mode'. Anyone who has experienced extreme circumstances where food, warmth and safety are limited and under threat will tell you that their focus is purely on the here and now. But it is something of a 'progress' myth to assume that simply because we do not have evidence of early humankind externalising an under-standing of existing in a moment in time, that he did not have it. The development of cave art, from as early as 45,000 years ago but increas-ingly common from 35,000 years ago, perhaps demonstrates a concept of a more distant past and future. If you frequented a cave with pre-existing rock paintings, your thoughts might naturally turn to the ancestors who made them before you; and if you added your own marks to the walls, you might contemplate the generations that would see them after you were gone. But there is no way to know where the genesis of our shared time lies. The appearance of grave goods later, about 13,000–15,000 years ago, gives us more conclusive evidence of a belief in a time beyond our own. Burying loved ones with their treas-ured possessions – a favourite knife, jewellery, an infant's toy – implies that these objects might be required in a future afterlife.

A few years ago, archaeologists discovered a 23,000-year-old human camp on the fertile shores of the Sea of Galilee in Israel. They found 140 different species of plant there, including emmer, barley, oats and, crucially, the remnants of lots and lots of weeds (weeds flourish in disturbed soil and cultivated land, which is why they are the bane of every gardener's life).* This site is the earliest known evidence of basic agriculture, some 11,000 years earlier than previously believed.

---

* They also found a stone grinding slab and sickle blades, indicating that the cere-als were being grown, harvested and processed in an organised manner.

These pioneering farmers were probably also observing the position of the sun, the phases of the moon and the migration of animals. Above all, they clearly had a conception of the future: they understood that if they planted something in the present, they could reap their rewards months later.

This is still some way from the modern experience of time as defined by the hours on a clock dial. For our ancestors, time was divided not by abstract numbers but by natural *events*, such as seasons and their related weather conditions. This is how the Kenyan-born philosopher Rev. Dr John S. Mbiti described event-based time in relation to traditional African hunter-gatherer communities: 'There is the "hot" month, the month of the first rains, the weeding month, the beans month, the hunting month, etc. It doesn't matter whether the "hunting month" lasts twenty-five or thirty-five days: the event of hunting is what matters much more than the mathematical length of the month.' Cycles of a longer duration, like a year, would be measured by the repetition of the agricultural cycle, such as the passage of two wet and two dry seasons, with these four seasons making one year. The exact number of days in a year was not important, 'since a year is not reckoned in terms of mathematical days but in terms of events. Therefore one year might have 350 days while another has 390 days. The years may, and often do, differ in their length according to days, but not in their seasons and other regular events.' In many ways this system makes better sense than our attempts to bend nature's unpredictable patterns to our will. To pin one's hopes on a natural event occurring on a numerical day or hour of a human-constructed calendrical system is to be doomed to disappointment.

Storytelling also played an important role in event-based systems of recording time. Without a numerical calendar to reference, tales about ancestors – and their experiences of good and bad harvests, floods, droughts, eclipses – were invaluable ways by which history acquired its shape, how the past informed the present and predicted the future. For some coastal Aboriginal communities living in modern-day Australia, these stories go as far back as the rising of the ocean at the end of the last Ice Age 10,000 years ago. Māori culture likewise places supreme value on genealogy and ancestry – on all that went before – and uses the wonderful word 'whakapapa' (pronounced 'fakapapa') to describe it. For them, a meaningful future is unimaginable without knowledge of the past.

Nature still influences our relationship with time – even in our increasingly digital age. British Summer Time, the act of adjusting our clocks backwards or forwards by an hour every six months to increase the amount of daylight we have in the winter mornings, shows that the light, rather than the hour, remains the decisive factor in getting up in the morning. We still calculate the length of a pregnancy according to lunar months (ten lunar months is forty weeks) or end our day on the beach by judging the movement of the tide. The changing colours of the leaves or a sudden chill in the air tell us summer is over far more viscerally than any date on a calendar.

What's more, our timekeeping is *still* event- and story-based. We say, 'That was just before you were born'; 'It was the summer after my GCSEs'; 'It was the month after our wedding', locating things around milestone moments in our own lives. Current generations will for many years think of things as 'Before' or 'After Covid' – an almost universal mass event, even though for those stuck at home during the pandemic lockdowns time lost all distinction. As the milestones that might have marked the year – weddings and holidays, parties and exams, even Christmas – were all cancelled, the days felt oddly 'out of time'.

Close your eyes and think of a watch.

I suspect you're picturing an analogue clock face, with the dial divided into twelve.* Two hands are rotating 'clockwise'. And the whole thing is mounted on a wearable strap.

---

* Although this perhaps depends on your age! As mobile phones and computers have taken over as our go-to devices for telling the time, analogue dials are becoming less common in public spaces. Many train stations now use digital displays. The lack of common use of analogue timekeeping has meant many school classrooms now have digital clocks.

All of these elements were established in the ancient world. And all of them were arrived at through dialogue with nature. The Sumerians, the first known Mesopotamian civilisation (located in modern-day Iraq and Syria), are often credited with inventing the first numerical system for measuring time. They developed the first written number system based around the number sixty, which still dictates how we quantify minutes, hours, angles and geographical coordinates. This number was easily divisible without complicated fractions or decimals. It was also divisible by three – helpful because most humans have an inbuilt body calculator for the three times table. Each of our fingers has three joints or knuckles, so one hand (not counting thumbs!) accounts for twelve finger knuckles; together, both hands total twenty-four. This counting system may well be the origin of the twenty-four-hour day.

A thousand miles west of Sumer, in ancient Egypt, scholars started to use the sun and stars to divide time even further. The name of the ancient Egyptian god of the sky, Horus, whose right eye was believed to be the sun, is the origin of the modern word 'hour'. About 5,000 years ago, Egyptians discovered that the Earth's solar year – the time it takes our planet to revolve round the sun – influenced the rising waters of the Nile, and coincided with the summer solstice and the prominence of Sirius, the dog star, in the night sky.

Our very idea of 'clockwise' is also a function of the sun, as well as an accident of location. The civilisations which shaped contemporary timekeeping systems were generally located in the northern hemisphere. And if you want to follow the path of the sun across the heavens in the northern hemisphere, you have to face south. From that position, the sun moves from left to right over the course of the day, with the countering shadow it casts creeping around from right to left: in other words, clockwise. This simple observation surely led our ancestors to gauge the time from the lengths and angles of the shadows cast by the people, buildings or trees around them. The sundial, the first clock 'dial' as we know it, was an attempt to harness this very phenomenon, replacing random shadow-casting objects with a designed vertical rod or shape called a gnomon.

No one knows who first invented the sundial or shadow clock. They appear all over the world, from the Stone Circle at Stonehenge (c. 3000 BC) in England, positioned to align with the sun on the summer and winter solstices, to the painted sticks used to make calculations from shadows at the ancient astronomical site of Taosi in China (c. 2300 BC). In the ancient Egyptian burial site the Valley of the Kings, the divisions of a very early sundial were found etched into a flat sheet of limestone on the floor of a worker's hut dating to the middle of the second millennium BC. The gnomon, a separate vertical stick pushed into a hole at the centre of the dial, has been lost but would once have cast its shadow on a semicircle drawn in black and divided into twelve sections approximately 15 degrees apart. The rough divisions were enough to allow its owner to pinpoint the start of the working day, lunch, and time to pack up and head home before it got dark. It is this pairing of gnomon and dial that creates a 'true' sundial.

Sundials served another important purpose: they were a community's focal point. Often planted in the heart of towns and cities, they provided the local population with a shared sense of time – one that everyone could access and work to. This collective understanding of time proved crucial to the development of civilisation. By charting the skies and measuring the movements of the sun, we were able to divide the lives and routines of large groups of people into ever smaller and more accurate parcels of time. These divisions made it increasingly easy to work together and schedule our interactions with others – whether it was farming, trade, education or governance – and in turn help us make plans for the future.

In the hours of darkness, the ancient Egyptians looked to the stars, using them like a vast celestial clock face (we still use zodiac and star groups to measure the passage of time).* Astronomers identified at least forty-three different patterns that included sꜣḥ (or Sah in English,

---

* In the night sky the constellation of the Great Bear, Ursa Major, marks the seasons clockwise. The bear's tail points eastward in the spring, to the south in the summer and northward in winter.

which includes parts of Orion), 'ryt (the transliteration of 'the jaws', modern-day Cassiopeia), knmt (possibly meaning cow and signifying Canis Major) and nwt (the Milky Way, and sky symbol of the goddess Nut). They also knew of the planets Mercury, Venus, Mars, Saturn and Jupiter; and they could calculate and predict lunar eclipses. Celestial calendars played an important role in planning the annual lunar festival, where swine were offered to the moon god and Osiris, the god of agriculture, at the season of the new moon.

When I imagine a watch, I always hear it ticking; a constant, nagging reminder of how fast our time on this planet is passing. Many earlier timekeepers likewise recorded the *passage* of time. Water clocks did this by harnessing the regular pace at which water flows through a hole. The earliest examples were surprisingly simple: they were essentially an earthenware vessel filled with a certain measure of water that then flowed into a second earthenware vessel. They relied on an accurate understanding of volume and flow rates. For these clocks, time would literally run out and need to then be refilled by a dutiful attendant. Alabaster and black basalt water clocks were used by the ancient Egyptians for this purpose, while Bronze Age clay items have been found on the coast of the Black Sea in modern-day Ukraine. Variations on this basic system developed all over the world, from ancient Babylon and Persia to India, China and Native North America and ancient Rome. In ancient Greece, *clepsydra* (meaning 'water thief') were used in the Athenian law courts to mark the time given to each speaker. Some of these clocks could even sound an alarm. In the design of a water clock dating from 427 BC, invented by Plato, a series of four vertically stacked ceramic urns is used that allows water in the top vessel to flow at a slow and controlled rate into the next vessel directly below it. When that urn is full, at a precise timed moment calibrated by its size and the speed of the flowing water that fills it, water floods through a siphon into the third urn below all at once. The sudden rush of water forces the air in the third vessel through a pipe near its top that whistles to sound the alarm. The fourth urn at the bottom of the stack collects the water ready for it to be reused.

In ninth-century England, King Alfred the Great of the West Saxons used candle clocks like a modern-day productivity guru, keeping a strict daily schedule consisting of eight hours for work, eight hours for study and eight hours for sleep.* His 'clock' consisted of six candles of uniform width and height. Each one took four hours to burn and was marked with twelve equally spaced divisions, each representing twenty minutes. So two candles marked the duration of Alfred's daily reading and writing (he was a passionate scholar who translated a number of Latin religious texts into Old English), and another pair stood guard as he planned battle tactics to defend his lands against invading Viking armies or mediated disputes between his subjects. The final two watched over the king as he slept.

As travel increased and people needed to tell the time on the go, many of these traditional timekeepers proved impractical: sundials were too static, clepsydras sloshed all over the place, and candle clocks were extinguished by the wind. In the second half of the Middle Ages, the hour- or sandglass were increasingly used alongside them. By the end of the thirteenth century, 'sand-clocks' were being used on ships. In his *Documenti d'Amore*, written between 1306 and 1313, Francesco da Barberino insisted that 'in addition to a lodestone, skilled helmsmen, good lookout, and chart, the sailor must have his sand-clock.' In the late fifteenth century, Christopher Columbus is said to have used a half-hour *ampoletta* (meaning ampoule, a kind of glass container), which was maintained by the helmsman and corrected using the midday sun as a reference. Sand-clocks helped sailors locate themselves not only in time, but also in space: by knowing *how long* it had been since you set sail, and the speed you were travelling (measured literally in knots, by extending a line with equally spaced knots overboard and timing how quickly the knots dragged out to sea), you could calculate roughly where

---

* Candle clocks, made from whale fat, probably originated in China around 200 BC. Their relatively stable and consistent burning rate made them useful indoors and at night.

you were and when you would reach land. This process is called dead reckoning and for centuries the sand-clock was the best device available. It would take another 500 years and a revolution in science and engineering before the mechanical clock could match the accuracy of the sandglass for the measurement of longitude at sea.

In the sixteenth century sundials became small and portable. Ring dials* (the smallest were about the size of a man's wedding band) were engraved metal rings that could be held up to the sun and read as the light shone through a small hole in the main band and cast a bright dot on a scale inside. The sides of the ring, formed from separate pieces of metal, could be rotated to adjust for the correct month and latitude to give an accurate reading. Their invention is credited to the sixteenth-century Dutch mathematician and philosopher Gemma Frisius (1508–1555), who in 1534 took his idea for an 'astronomer's ring' to the engraver and goldsmith Gaspard van der Heyden – a collaboration of science and craftsmanship that foreshadowed the watchmaker's art.

The wrist-strap we imagined earlier is a defining feature of the watch, because it makes time wearable and therefore personal. The ring dial was significant for the same reason. This was the first timekeeper that could be tucked in a pocket or suspended on a cord or chain and carried throughout the day. Petite, lightweight, and completely unaffected by the motion of their wearer, ring dials proved so practical that they were used for several centuries after the invention of the watch. They even get a cameo in Shakespeare's *As You Like It*, when Jacques describes meeting a fool in the forest who makes a big show of pulling 'a dial' out of his pocket, telling him 'very wisely "It is ten o'clock."'

That passage reminds me of *my* foolish moment in the forest, as I looked up at the geese and obsessed over my own prized timepiece.

---

* Properly termed *universal equinoctial ring dials*.

In the end, the deadline I was working to during that first winter in the countryside passed me by, just like the birds: I didn't finish the watch for another three years.

I always find it comforting to remember that however mechanised and digitised our experience of time seems today, it will always be underpinned by natural forces that remain completely out of our control. And that, in the end, some things still take as long as they take.

2

Ingenious Devices

'Measure what is measurable, and make measurable what is not so'
Attributed to Galileo (1564–1642)

W hen I was seventeen I dropped out of school and wound up on a silversmithing and jewellery course. As far as my friends and teachers were concerned, I might as well have run away to join the circus. I had always loved art, but I had taken all science A levels in pursuit of my dream of becoming a pathologist. School wasn't encouraging. My careers adviser implied that medicine was not really for kids from my working-class background. And though I loved science, the way it was taught felt so rigid, dry and cold. So little of it was hands-on. I'd spent all year looking forward to dissecting a sheep's heart in biology only to discover the exercise was being cancelled after another student fainted. Halfway through the course, I had my moment of rebellion: I decided if science wasn't going to have me, I'd run away to art school.

The tutor on the silversmithing course was a master goldsmith from Austria who had started his apprenticeship at thirteen and was due to retire the year I finished the course. Peter had such a wealth of knowledge that he was humbling to be around. I still adhere to many of his creative philosophies. He is the reason I only make watch cases out of precious metal. One day he saw me begin to make up a design with gilding metal (a copper-based alloy with similar working properties to 9ct gold but without the price tag) and invited me over to the workshop safe. I explained I didn't have the money to buy precious metals, so he sorted through trays of gold sheet and wire and gave me what I needed. 'Rebecca,' he said, 'you have taken a long time working on that design, and it is a beautiful design. You must only work in materials worthy of your efforts. Don't worry about the metal costs, we'll sort it out another time.' And so I made the piece in gold, a brooch in the form of a phoenix with black diamond eyes, blood-red rubies across its breast and a diamond in its tail. I pierced the individual feathers in its wings to create tiny windows, or cells, through the metal that I filled with glass enamel. He was right. It's daunting, not to mention expensive, to work with precious materials, but you could spend a whole year making something from gilding metal that would hold little value on completion. If you have the confidence, and capital, to make something in precious metal, its value will only increase. Even so, I ended up scrapping the brooch a few years later, during one of many desperate moments teetering on the brink of financial ruin. I still remember crying as I twisted it apart with pliers to break the stones from their settings.

Peter also taught me how to make mistakes. I was making my first solitaire ring with a rex claw setting – one of the most traditional single-stone engagement ring designs. I'd got a bit carried away filing the claws that would hold the stone in place and accidentally made them too short. Deflated, I asked Peter if there was a way the setting could be saved. He sat at my bench and proceeded to re-dress the claws, making them perfectly usable again. I remember calling him a genius and thanking him profusely for saving me a week of work. His response? 'Rebecca [he always started with "Rebecca . . ."], do you know how I know how to fix this? I know because I have made this mistake myself and learned from it. It's okay to make mistakes – as long as you learn from them.' I still think of those words most days.

Peter's jewellery course was where I learned metalworking skills like soldering and saw piercing – the art of using a fine saw blade just a millimetre or so in depth to cut intricate shapes. I was still interested in science and engineering and, as I became more confident, I started to incorporate hinges, pivots and other simple devices to introduce movement into my jewellery. As a visual thinker, being able to see how things work out in the real world made sense to me. In retrospect, that's how I've always learned. By experimenting, watching and testing the outcomes of my physical interactions with things.

As time went on, I started experimenting with basic automata – moving mechanics that imitate living things. I've always loved orreries – clockwork models of the solar system. To me they're some of the finest examples of nature as represented in mechanics: our very human way of containing the universe in a device small enough to be examined on your desk. So, for my final project in jewellery and silversmithing I designed an orrery where each of the planets was a removable and wearable piece of jewellery. Saturn was a pendant whose rings could turn independently of each other and in different directions. The sun was a ring made up of spiky 'flames' that spun and appeared to flicker. It would also have made a good knuckleduster in a fight.

At that stage I lacked sufficient understanding of the mechanics I needed to create a properly functional orrery that could chart the movements of the planets. I was also up against a tight deadline. I took enormous liberties with the planetary positions, and the rings that held them had to be rotated independently by hand, rather than an interconnected system powered by a motor or turning a handle. The real thing didn't live up to my design; it was nowhere near as advanced as I wanted it to be. I needed more time and knowledge to fully realise it, but I was nonetheless captivated by the workings of the interconnected moving parts.

During the end-of-year show, my orrery caught the attention of the horology students. A small group (one of them was Craig) made a beeline for me, seeing that I shared their interest in making small fiddly things that moved. Up until then, when I thought of watchmakers, if I thought of them at all, I'd imagined people changing batteries or straps in shopping centres. But in their workshop, surrounded for the first time by the whirring lathes and mills, and the smell of metal and swarf, I realised watchmaking could allow me to be an artist, designer, engineer and physicist all at once. By the end of the course I had joined them.

I wish I still had my orrery. But, as with the phoenix brooch, I ended up scrapping it to pay the rent.

My own early, fumbling attempts to make things move always play on my mind when I think about the earliest clockmakers. Without constant human maintenance, water, sand and candle clocks were little more than early egg-timers. 'True' clocks and watches require a self-renewing or mechanised power source. Although mechanical clocks and watches didn't appear until the eleventh and sixteenth centuries respectively, the developments needed to make this leap were already under way over a millennium earlier. In ancient Mesopotamia, all kinds of engineering advances – from hydraulic machinery, crop irrigation systems and textile mills to mass production of bricks and pottery, wheeled chariots and even the plough – supplied the groundwork for mechanised timekeepers.

I like to imagine the astonishment that must have greeted the first hydromechanical timekeeper that appeared in Europe. In 802 AD one of King Charlemagne's favourite diplomats returned from a mission to Baghdad with a shower of gifts designed to dazzle and impress. They had been sent by Caliph Harun al-Rashid and epitomised the Islamic Golden Age that had flourished under his rule. While the star of the show was undoubtedly Abul-Abbas, the adult Asian elephant, who must have caused mayhem as he plodded through the streets of Aix-la-Chapelle,* Harun also gave Charlemagne a brass water clock with a mechanism capable of striking the hours. Witnesses described how at the end of each hour, a cymbal would clang while a model horseman emerged from one of twelve doors. We don't know exactly how the mechanism worked, but it would have been a system of weights and ropes both powered and controlled by the changing water level as it seeped out of a hole. For the lucky Europeans who beheld it, this clock would have seemed like magic.

* Now Aachen, in western Germany.

In the eleventh century, the Spanish Muslim astronomer and inventor of scientific instruments, al-Zarqali, designed and created a water clock capable not only of telling the hours of the day, but also of displaying celestial information. Located in the ancient city of Toledo in central Spain, the clock was renowned for its ability to illustrate the current phase of the moon, using two basins that would gradually fill with water and then empty over the course of twenty-nine days, to echo the waxing and waning of the moon. Al-Zarqali managed the ever-changing water levels with a subterranean pipe system that compensated for water removed or added to the basins, should anyone attempt to interfere with them. I can just imagine a curious child (it would have been me) sneaking off to al-Zarqali's clock while their parents bargained with a nearby market stallholder. After removing some of its water to see what happened, they would then have watched in wonder as the dish magically refilled itself to exactly the right point.

My home city of Birmingham has a water feature at its heart, too. Designed by Indian sculptor Dhruva Mistry in 1992, a fountain called *The River* takes pride of place outside the town hall in Victoria Square. At the top, a giant reclining female nude, cast in bronze, holds a pitcher of water that flows into a palatial upper pool where it is retained until it overflows its lower edge and runs down a series of steps that empty into a second large pool at the bottom. Affectionately referred to by locals as the 'Floozie in the Jacuzzi', she provides a convenient meeting point that everyone knows. Some people offer her coins and make wishes, and, on a hot summer's day, a few even venture into the water (albeit illicitly) to splash around and cool down. Installations situated in public spaces invite people to interact with them.* These are shared spaces, over which we all have some sense of communal ownership. I can imagine a curious child playing with the water in al-Zarqali's clock, making the most of their parents' distraction as they bargain with a nearby market-stall holder,

---

* Sometimes the interaction is a little too much. After a few years out of action, during which time the water was replaced with a flowerbed, the 'Floozie' was restored to its watery glory in 2022. But within a month it had been damaged by someone adding bubble bath – not a problem al-Zarqali ever faced.

by removing some with their cupped hands to see what happens, then watching the magic unfold as, no matter how much they remove, the dish mysteriously refills itself to exactly the right point.

Al-Zarqali's clock remained in Toledo until a later inventor, intrigued by its construction, was permitted to dismantle it for examination but apparently proved unable to put it back together again. I've always been moved by that story because I've worked on many timepieces that have had the same fate. The initial conversation with the owner usually starts with them saying, 'Well, this is a bit awkward, but I . . . and then I . . . so you see . . . cutting a long story short . . .' before admitting that their inherited grandparent's/bought at auction/gift for a significant occasion watch is now a bag of bits that they have no idea how to put back together. Curiosity has been killing clocks and watches for centuries.

Nearly 4,000 miles east of Toledo in the Henan province of China, in 1088, the astronomer Su Song was commissioned by the emperor to create the world's finest water clock, intended to showcase a showpiece for the intellectual prowess of the Song dynasty. The brief for Su Song's clepsydra involved a number of complex celestial displays: at the time, dynastic houses were ruled according to a heavenly mandate, or *tianming*, which required an ability to track and predict astronomical events for interpretation, as a guide to inform bureaucratic decisions. With its bronze astronomical and armillary spheres, and automata mannequins beating gongs at important hours of the day, Su Song's clock didn't only confirm China's technological pre-eminence, but functioned as 'a hotline to divinity, a conduit through which heavenly wisdom flowed into the Imperial court'.

Starting with a scaled-down model prototype he'd made from wood designed in the style of a pagoda tower, Su Song and his team of craftsmen and engineers spent eight years building his hydromechanical celestial clepsydra. The finished clock was 40 feet high – around the height of a four-storey building – and was powered by a giant water wheel, 11 feet in diameter, with thirty-six buckets around its rim. As each bucket filled with water and became heavy enough to trip the mechanism, it would fall forward, rotating the wheel and placing the next empty bucket in the path of the flowing water supplied from a separate tank to keep the volume consistent.

What fascinates me, as a watchmaker, about Su Song's pioneering piece of engineering is that it represents the first ever escapement – the mechanism that alternately checks and releases the power of the gear wheels, giving the clock the potential for an indefinite duration. (Infinite, that is, so long as there were human monitors constantly available to maintain the water level in the supply tank and perform any necessary adjustments and servicing.) This group of components capable of locking and releasing motive energy from a source like water, gravity or a spring would become instrumental in the invention of the first fully mechanical clocks. It was also, I should add, the first moment in history when the 'tick, tock' sound of a clock was heard.

All of these early mechanical devices were suffused with the sheer joy of experimentation and discovery, the result of trial, error and endless possibility. At the turn of the thirteenth century, Ismail al-Jazari, a Muslim polymath, scholar and inventor from Upper Mesopotamia, took the development of mechanics to an entirely new level. The chief engineer of the Artuklu Palace in Turkey, al-Jazari is sometimes referred to as one of the 'fathers of robotics'; he was a master of automata. His *Book of Knowledge of Ingenious Mechanical Devices*, rumoured to have inspired Leonardo da Vinci some 250 years later, details around 100 mechanical inventions accompanied by delightfully vivid hand-painted illustrations. They include automata peacocks, a humanoid waitress powered by water who could serve drinks at parties, a band of musicians who 'played' music to party guests, and several complex candle and water clocks.

Where Harun sent a clock accompanied by an elephant, al-Jazari went one better and designed a water clock that took the form of an elephant. On the creature's back, perched on a Persian carpet, an Arabian scribe with a pen in his hand sits in a golden howdah. A mahout, or elephant trainer, rides at the front, and red Chinese dragons and an Egyptian phoenix adorn the top. Concealed in the elephant's belly is a water-filled basin. Floating on the water is a bowl, punctured with a small hole, attached by strings to a pulley mechanism. The bowl slowly fills with water and sinks down, causing the scribe to rotate once every minute. After half an hour the bowl fills completely and sinks to the bottom of the basin, triggering a see-saw mechanism that releases a ball into the mouth of one of the dragons. The weight of the ball causes the dragon to tip forwards, pulling the sunken bowl back up to the surface of the water below. This process also triggers a human statue on top of the carriage to raise his hand. A cymbal sounds every half-hour, the phoenix spins and the mahout moves his beaters. Once the cycle is complete, and the performance ends, the characters return to their original positions and wait for the bowl to fill again.

The elements of the clock were deliberately designed to embrace the combined global knowledge of engineering at this point. According to al-Jazari, 'the elephant represents the Indian and African cultures, the two dragons represent ancient Chinese culture, the phoenix represents Persian culture, the water work represents ancient Greek culture, and the turban represents Islamic culture'. The elephant clock continues to amaze today. In 2005 a monumental replica was built as the centrepiece of a shopping mall in Dubai. Sitting in the middle of a vaulted marble hall, surrounded by eager shoppers taking photographs, the Dubai elephant clock is once again a focal point of shared time.

When it came to timekeeping, medieval Europeans lagged some way behind their Chinese and Islamic counterparts. But as Europe entered the Renaissance, a growing number of astronomers emerged, within the Catholic Church and more generally from elite society, to push clockmaking into a thrilling new era. Deeming water an unreliable power source – in European summers it tended to evaporate, and in the winters it froze – they made major advances in the development of genuinely mechanical clocks.

The key to making a fully mechanical and comparatively low-maintenance timekeeper was finding a reliable power source. This problem was eventually solved with a little help from gravity. Some time in the fourteenth century – we don't know exactly when, or who invented it and where – a remarkable clock appeared. It was powered by a rope attached to a heavy weight at one end fixed to a horizontal arbor (a bit like the spool you wind thread into) and the mechanism at the other.* The oldest surviving clock with such a mechanism was constructed in 1386 and now lives in Salisbury Cathedral in England. In this case, power was supplied by two stone weights; as the weights descend, ropes unwind from large wooden spools. One spool drives the going train (responsible for keeping the time) while the other drives the striking train (which chimed the bell to give the time signal). The arbor could be wound manually with a crank, slowly coiling the rope around it and raising the weight upwards. Ratchet work allowed the arbor to turn one way but locked it, a bit like reeling in a fish on a line, preventing it from spinning back the other way. Once fully coiled, or wound, the force of gravity would then pull the weight down, causing the arbor to want to turn back on itself at an uncontrolled speed. So a new device to control the speed of the rotation was required: we call it a *verge escapement*.

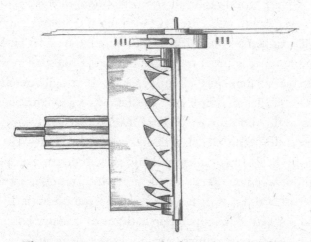

The verge escapement was the first to be used in fully
mechanical timekeepers and remained in use
until the early nineteenth century.

---

* There is a detailed account of another turret clock in Chioggia, near Venice, which also dates back to at least 1386 and survives – albeit in a currently decommissioned state.

The verge in these early clocks is made from a long, thin rod of steel (the staff) topped with a horizontal bar, making it look like a capital T. At the top and bottom of the staff are two rectangular 'flags' attached at about 90 degrees to each other. These flags are spaced so that when one of them swings with the oscillating staff, it catches a tooth of the crown wheel (so called on account of its jagged teeth, which echo the shape of a crown). As the wheel turns forward, one flag swings to allow a tooth through, while the other flag catches another tooth, thus controlling the release of power. Once the flag has made contact with the crown wheel, the momentum impels the verge to swing back the other way, allowing the other flag to catch a tooth while this time the first flag lets a tooth through, and so on, back and forth. This cycle repeats over and over, thousands of times every hour, controlling the release of power – and creating an audible *clunk, clunk, clunk* each time a flag catches a tooth.

The verge escapement made possible the first church clocks – devices whose workings lived in turrets and towers, and which loomed over towns and cities to be visible for miles around. The earliest clocks had no dials; rather, they struck the hours on a bell to indicate the time, in order to guide public routines such as worship. The Elizabethan dramatist Thomas Dekker described how the clock bell could be 'heard a farre off, whither we lye in our bed in the night, or in the day time we be farre from a Dial'. The word 'clock' itself derives from the Medieval Latin *clocca*, and French *cloche*, which both mean 'bell'. In late medieval and early modern Europe, right up until the seventeenth century, time was still a public rather than private matter and was delivered, literally, from on high. In an age when 90 per cent of Europe's population was peasantry, the idea of a personal clock with which you might self-determine your time was still a long way off.

Grand public turret clocks are about as far removed as you can get from the microscopic world of the watchmaker. A few years ago I had the pleasure of visiting the workshops of Smith of Derby, a turret clock restoration and manufacturing company that was founded in 1856 and is now managed by the fifth generation of the Smith family line. I imagine the experience would be quite surreal for most people but, coming from my own workshop, I felt like a Borrower who had just emerged from behind a skirting board into a realm of giants. While the tools we work with are fairly similar, Smith's are five, ten or twenty times the size of mine. It felt incredibly familiar, and yet completely different. When I fit the hands to a watch dial (one of the last things I do when I put a watch back together) they are often no longer than the tip of my little finger. In the workshops of turret clockmakers, giant hands, some taller than you or me, have to be winched back into place on dials the size of a double-decker bus.*

Aside from differences in scale, turret clockmakers face struggles that are almost inconceivable to watchmakers. They battle gales and freezing cold weather up church towers, swing in harnesses suspended off roofs, scrape away piles of acidic pigeon poo that have clogged up the mechanism, and occasionally face altercations with angry nesting seagulls. All of this makes me feel grateful to be working in a warm, safe workshop. But there is nevertheless something magical about turret clock movements. Their aesthetic is pure H.G. Wells. They're curious science-fictiony machines that clunk and clang and whir. The action of the movement is heavy and methodical, almost symbolic of the way 'clock time' would go on to dominate the modern world.

The verge escapement was not, in truth, very accurate – some clocks lost or gained several hours over the course of a week – but it

---

* To give you an idea of scale, the largest clock dial in existence can be found at the top of the Abraj Al Bait Towers in Saudi Arabia. Constructed in 2012, it measures 43 metres across – just short of the length of two male blue whales lined up nose to tail.

provided a foundation on which later inventors and engineers could build, making timekeepers that were more complicated, more impressive and more eye-catching than people only a century earlier could have imagined. Medieval church clocks were soon displaying a highly sophisticated range of information, including planetary positions, predicted eclipses, lunar phases and high and low tide times – often brought to life with elaborate automata. Financial records from Norwich Cathedral between 1321 and 1325 describe the commission and installation of a mechanical clock whose displays included a model of the sun and moon and fifty-nine moving sculptures carved from wood, including a choir and procession of monks.* In the fourteenth and fifteenth centuries, the Church saw these complicated astronomical clocks as grand public representations of Christian cosmology, designed to illustrate, through ornament and automata, the meaning of the passage of time to followers of the faith.

Astronomy was central to the worldview of medieval Christians, who saw God as the divine architect of the universe. Pictorial depictions from the period show God as a geometer, dividing compasses in hand, mapping out his plan for the cosmos. Celestial events were also believed to have a direct effect on human life. Marriages, diplomatic decisions, even surgeries were performed with reference to the alignment of the moon and stars. Each of the twelve zodiac signs had an associated part of the anatomy and it was held to be dangerous to operate on a part of the body if the moon was in its associated zodiac sign. Devices like lunar volvelles could be consulted to calculate the position of the moon in relation to the sun, which when checked against a zodiac man – an illustration of a man's body on

---

* The inventories describe in detail the use of brightly coloured paints and gilding. The whole project required the talents of a wide range of skilled craftspeople, who were employed on the build for three years – from blacksmiths and carpenters to masons, plasterers and bell founders. The total cost of the clock was listed as £52, which at that time, had it been a one-man job, would have covered the wage of a single skilled master craftsperson for over fourteen years of solid work.

which the anatomical parts associated with each zodiac sign were marked – would tell you if the signs were auspicious. If you had an ingrown toenail and the volvelle said the moon was in Pisces (the sign associated with feet) then bad luck; you'd have to wait a month before you could have the nail removed.

Just as in ancient Egypt, where the study of horology and astronomy was largely conducted by priests, so in medieval Europe monks were among the fortunate few who, in the absence of worldly distractions and without having to worry about keeping a roof over their heads, were able to dedicate significant time to the furtherment of knowledge. A number of the earliest clockmakers were men of the church, including the monk and natural philosopher Richard of Wallingford (c. 1292–1336) who designed an astronomical clock in the 1320s, and the clergyman and astronomer Jean Fusoris (c. 1365–1436), the designer of a monumental astrological clock for Bourges Cathedral in France.

Astronomers drove clocks towards ever greater accuracy. In order to make measurements of an observed phenomenon, like a lunar eclipse or a passing comet, they needed a timekeeper that was capable of measuring perfectly equal time increments. Ancient methods of division often used the separation of day and night to mark the passing of each day, creating hours of varying length depending on the time of year and the distance between dawn and dusk. Mechanical clocks, however, relied on gearing to control the motion of the hand, meaning that (providing the clock was running correctly and didn't stop) there was no variance in how long it took each hand to make a full rotation. The regimented nature of a mechanical clock made a totally controlled uniformity possible.

Galileo, who is rumoured to have said 'measure what is measurable, and make measurable what is not so', is credited with discovering one of the most significant improvements in accuracy: pendulum isochronism (how a pendulum, in the absence of variables such as wind, will swing at a constant rate). The story goes that, while attending mass in Pisa Cathedral, the nineteen-year-old Galileo looked up and noticed the regular, repeated swinging of an altar lamp suspended from the ceiling. In that moment, it struck him that this swing could be used to trigger the regular release of power from a mechanism. The idea rumbled away at the back of his mind for years, until in 1637 he designed the first ever pendulum-regulated clock, which used a swinging weight to trigger the release of the escapement. Galileo died five years later, and never saw his ingenious concept become a reality. It would be another fifteen years before the Dutch physicist and mathematician Christiaan Huygens converted it into a working clock mechanism.

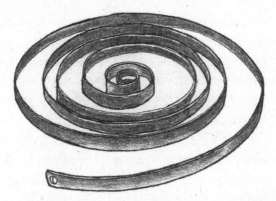

A typical traditional mainspring. This form was used from its advent until the twentieth century when its shape was refined to generate a more uniform torque as it unwound.

But if astronomical observations were to be made in a range of locations, clocks needed to be portable. The most significant advance to aid this process was the introduction of something called a mainspring, in which a tightly coiled steel spring replaced weights as the power source. We can't be sure who invented the mainspring, but it's highly likely the technology also came out of locksmithing and emerged from northern Italy. The earliest surviving spring-driven clock dates to 1430. These springs, which were long but wafer thin – a bit like a ribbon you would tie round a gift – were coiled into a drum or barrel, and wound around a central arbor using a crank or key. The spring coils like a tight spiral round the arbor, then, when released, pulls back as its elasticity forces it to unravel, dragging its outer hooking with it to create the all-important rotating motion. To prevent the mainspring from suddenly releasing all of its power through the train wheels at once, the speed of this rotation is controlled in exactly the same way as the earlier weight drive was, via an escapement.

Mainsprings allowed clocks for the first time to become independent of gravity and thus small enough to move around with us on our travels. If religion had supported the emergence of architectural mechanical clocks across Europe and scientists had made time-keepers more accurate and practical, the wealthy now rebranded these ingenious devices as status symbols. Over the course of the fifteenth century, clocks became a common sight in the homes of aristocrats and wealthy merchants – especially those with an interest in astronomy. They used these objects to show off their knowledge of cutting-edge technologies, just as we do today when we queue up for days outside an Apple store to buy the latest iPhone. Their cost, and therefore exclusivity, made them objects of desire, and soon materials like gilded brass and copper replaced the iron used in the earliest mechanisms. As engravers and gilders became involved, clocks grew ever more ornate.

The British Museum owns a remarkable clock, made in 1585 by Augsburg clockmaker Hans Schlottheim, probably for the Holy Roman Emperor Rudolf II. The clock, which is mounted on a gilded brass galleon, was designed to 'sail' down the middle of a busy banqueting table 'firing' miniature smoking automata cannons while humanoid figures moved around on deck. The hours were chimed on bells suspended below the crow's nest while a mechanism in the hull played music to the beat of a drum. In the midst of this spectacular performance, the tiny clock dial on the ship's bridge is almost entirely missed.

In the sixteenth century, these extraordinary clocks weren't, in a way, all that extraordinary: many master craftsmen around Europe were making such mechanical marvels to satisfy an insatiable demand among the elite. There was just one problem: these elaborate machines could only be admired at home. They relied on their owner's friends, colleagues and clients to come over for dinner. By this stage Europe's ruling classes were ready for objects that, while equally intricate and curious, were portable enough to take around the world. For that to become a reality, the clock needed to be small enough to wear.

Small objects are good at disappearing. They can be stolen, destroyed or misplaced. They can vanish into the bowels of private collections. They can hide in the backs of drawers, in shoeboxes under beds, and even under floorboards. But then, sometimes, they are discovered again by chance.

So it was with the world's oldest known watch, which was found in a box of old clock parts at a London flea market in 1987 for £10. To the untrained eye it didn't look like a watch. It was essentially a ball, roughly the size and weight of a hen's egg, formed from two hemispheres of sheet copper, which had been hammered, or 'raised', to form a near-perfect sphere. On top of the ball is a hoop, through which would have run a chain, so it could be worn round the neck. Underneath it, three little feet allow it to sit on a table without rolling over. The case is decorated with a crude engraving of figures, a village scene and leaves. The upper segment is pierced with a number of comma-shaped holes through which one can just see the dial inside. To tell the time you unclip the top of the ball, which hinges back to reveal a single hour hand (the earliest watches weren't accurate enough to warrant a minute hand, and smaller increments of time weren't as important to their owners as they are now) circling a track of Roman numerals on an engraved dial. The watch is marked with the initials 'MDVPHN'. This offers us the first clue as to its origins: MDV is the date – 1505; PH is for Peter Henlein, a watchmaker known for making small portable mechanical clocks at the time; and N is for Nuremberg, where this one was made.

The buyer of this unusual object initially doubted its authenticity and sold it a few years later. Its next owner took the device to an expert, and they too were told that it was fake. The watch was sold again, for an undisclosed but presumably low sum. The third buyer then subjected the watch to detailed scientific analysis that proved, beyond all reasonable doubt, that the piece was indeed made in or around 1505 and was likely to be genuine, and therefore the earliest surviving watch in history. The watch is now estimated to be worth between £45 and £70 million.

Of all the remarkable facts about this little device, what's perhaps most remarkable is that we know something of its maker. We know that Peter Henlein was born in Nuremberg in 1485, the son of a brass worker, and was apprenticed as a locksmith, as were many of the first watchmakers. We also know that the decisive turning point in his early life happened not in a workshop, but in a tavern. In 1504, when he was nineteen, Henlein was involved in a brawl in which a fellow locksmith, Georg Glaser, was killed. As one of those accused of the murder, he begged for asylum at the Franciscan Monastery of Nuremberg and sheltered there from 1504 to 1508.

In the fifteenth and sixteenth centuries, the city of Nuremberg in southern Germany was one of Europe's creative and intellectual powerhouses. It was a centre of the German Renaissance, home to Johannes Gutenberg and his printing press (established in 1440), and Albrecht Dürer, who had set up his workshop there in 1495. Its monastery was also a magnet for academics and craftsmen, and doubtless provided Henlein with access to new tools and techniques as well as the work of visiting mathematicians and astronomers. Henlein had, serendipitously, found himself in an environment conducive to exploiting his talent.

It was perhaps in the monastery that Henlein learned to make the miniature fusee that he added to his watch. The fusee (pronounced fyu-zi), which connected to the mainspring, might have first been developed as a crossbow mechanism, and appeared in a design by Leonardo in 1490. As the watch is wound, the spring coils more tightly, storing energy, which then turns the barrel containing it as it uncoils – like the rotating dancer in a musical box or a mechanical egg-timer. And just as a dancer's pirouettes become slower and slower over time, the rotational power in the watch spring starts out strong but then diminishes as it unwinds. The helter-skelter-shaped tapered fusee helped regulate this, using a principle similar to a bicycle's gears.

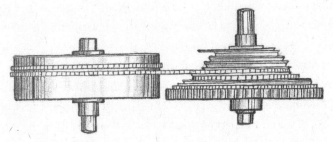

The fusee attached to a mainspring barrel by a fine bicycle-like chain. The earlier gut-line version worked in exactly the same way.

Henlein quickly developed a reputation for making clocks of extraordinary ingenuity and impeccable craftsmanship. He received commissions from the Nuremberg council for an astronomical device and made a public clock for the tower of Castle Lichtenau. But his speciality, it seems, was the creation of small ornate spherical watches, just like ours, that hung from chains as jewellery or as chatelaine brooches attached to clothing. In 1511, the scholar Johannes Cockläus described how Peter Henlein, 'out of a little iron constructs clocks with numerous wheels which can be wound up at will, having no pendulum, go for forty-eight hours, and strike, and can be carried in the purse as well as the pocket'.

It is harder to decipher the man behind these wonderful devices. As one of the few named watchmakers from that time, his character has repeatedly been mythologised. His modern fame is the result of Walter Harlan's play *The Nuremberg Egg* (1913), which was adapted into a film and released in 1939 as part of a Nazi propaganda drive to promote stories of German supremacy. The final cut was approved by Joseph Goebbels. The play and film make out Henlein to be a loving husband and dedicated artisan who dies from a weak heart. Later research, however, reveals a darker side. Analysis of this first watch movement has shown what appear to be the micro-engraved initials 'PH' marked over and over again on the metal of the mechanism, unreadable to the naked eye, which my psychologist friend suggests could indicate a sociopathic and narcissistic personality. We know that Henlein was capable of violence (the murder case against him was dropped after he paid blood money to the victim's family, rather than his innocence being proved), and that he offered fulsome support to his brother, Herman, who was beheaded for the murder of an eight-year-old beggar girl, in what is believed to have been a sexually motivated crime. Peter lacked any empathy for the murdered child and her family, and repeatedly attempted to have his brother pardoned. Put simply: Peter Henlein is one of the only celebrated watchmakers from history with whom I'd never want to go for a pint.

Henlein's watch is in some senses quite simplistic. The metals are raw – there is none of the advanced polishing or fine finishes associated with luxury modern watchmaking; the movements are made of iron – not an ideal material for a watch as, when struck, iron atoms tend to align themselves to the Earth's magnetic field, turning the movement into a magnet and disrupting the mechanism; and the engraving that adorns the surface is rustic and naive. And yet, as a modern-day watchmaker I am in awe of the skill that lies behind this device. Henlein's watch was made before high-quality magnification, digital measuring devices, motorised lathes or drills. Every element of his watch – every tooth of each wheel and pinion, and every last tiny screw – was made and assembled by hand. And, remarkably, five centuries later, watches like this one are still running, albeit with the care and attention of later restorers like me.

We have arrived at our first remarkable, mechanical, wearable timepiece. I see this watch as both a culmination and a beginning: the culmination of tens of thousands of years of a human journey towards personal, portable and mechanised time; and the beginning of a much faster-moving story – lasting little more than five centuries – about humans and machines.

3

Tempus Fugit

Time is more use to us than wealth or fate
Because it changes when appropriate.

Old age is an ill none can cure
And youth a good that nobody can store.
As soon as man is born his death is sure
And those who seem happy merely struggle more.

Mary, Queen of Scots, c.1580

At first sight the hollowed black sockets, gaping nasal cavity and fleshless rictus grin look ghoulish. But I try to set aside these initial impressions, to perform a kind of Etch-a-Sketch clearing of my mind, in order to imagine what this watch meant to its first owner. I'm looking at a sixteenth-century death skull watch, in the collection of the Worshipful Company of Clockmakers, which was once believed to have belonged to Mary, Queen of Scots.

The watch is silver, about the size of a small satsuma, and covered in delicate engravings. On the forehead of the skull, a sandglass- and scythe-brandishing skeleton stands with one foot at the door of a palace and the other at the door of a cottage – a reminder that Death haunts prince and pauper alike. On the back of the skull, the figure of Time – also holding a scythe – is flanked by a serpent eating his tail, and Ovid's classic phrase '*Tempus edax rerum*' ('Time devours all things'). The sides of the cranium are pierced by an intricate network of holes. Inside the skull lies a bell that the mechanism struck to sound the time. To read the time, you snap back the jaw to reveal a dial hidden in the roof of the mouth. The workings of the watch lie where, in a living skull, the brain would be.

This watch is a devotional object from a time when life was short and death an inescapable reality of everyday life. Merely to glance at it was to be reminded that at any moment you might be called to meet your maker. It's a sobering thought for any of us, but especially anyone who has ever brushed against the edges of their mortality.

I was raised a strict, practising atheist. We lived in a multicultural neighbourhood alongside Sikhs, Hindus, Muslims, Irish Catholics and Polish Jews, but my dad wanted nothing to do with his own Catholic upbringing. Dad – who had been a social security officer in nearby Handsworth before leaving work to bring up my sister and me – always told us that we should respect the religions of others while avoiding having one of our own. Our West Indian next-door neighbours were good family friends and, though they never discussed their Christian beliefs with us, they knew our leanings. Every so often they would gingerly

An example of an early-seventeenth-century silver skull watch.

post a flyer from their church through our letter box, inviting us to find the light of Jesus. I remember asking my dad why they did this when they knew we didn't believe in God. He explained that it was because they liked us, cared about us, and were worried that we were going to hell for not believing in God. It was a kind gesture, he said, and we should be grateful – if unconverted.

And yet – I've always been drawn to churches and cathedrals. Is it the peace that seems to penetrate the walls of worship, or the intense colours cast through a stained-glass window? Whatever it is, it's hard not to feel moved when you experience a full choir or organ recital in a sacred space. We can lose ourselves here, in time and in space, reminded of our own cosmic insignificance. Religion makes you feel small but at the same time part of something bigger. This was definitely true in the sixteenth century. A sense of belonging to God's universe shaped every aspect of a person's life back then, including attitudes to time – and therefore watches.

At the end of the eighteenth century a flurry of excitement spread among London's antique dealers when a watch believed to have belonged to Robert the Bruce, Scotland's Warrior King, was identified. This would have been a particularly impressive find given that Robert died in 1329, more than 150 years before the invention of the watch.

This story goes to illustrate that discovering the exact provenance of a watch is often tricky and distorted by hope. And Queen Mary's skull watch is no exception. Legend has it that she kept it with her at all times and only gave it to Mary Seton, her favourite lady-in-waiting, shortly before her execution.* But in the early 1980s, Cedric Jagger, former keeper of the Clockmakers' Company collection, discovered several watches that over the centuries had been claimed to be Mary's, in addition to a dozen copies. In all likelihood we'll never know whether any of the skull watches that have survived to this day were the one possessed by Mary – or if, indeed, her watch has survived at all.

But say our watch *was* Mary's, what would it have meant to her? It would have had monetary value, of course. For a queen, jewellery was a valuable commodity – a currency that could be used to fund

---

* Mary Seton, one of the famous 'Four Marys' who accompanied Mary, Queen of Scots from Scotland to France as a child, wasn't with her during her final imprisonment, so it is assumed that the watch was left to her servants along with other personal items such as jewellery, letters and small portraits, with instructions to distribute them when they themselves were eventually released.

wars, buy alliances or broker deals. It also probably had sentimental value. Mary was apparently given the watch by her first husband, the king of France, Francis II. Mary and Francis married in 1558, when they were just fifteen and fourteen years old respectively. Their marriage appears to have been exceptionally happy, albeit childlike, as the two had been raised together since childhood. But Francis, who had always suffered ill-health, died less than three years after their wedding and seventeen months into his reign, of an ear infection that led to an abscess in his brain. Mary lost her mother, whom she loved dearly, the same year. Grief-stricken, she had little reason to stay in France. In 1561 Mary returned to Scotland to reclaim her place on the throne, as a devout French Catholic in a now-Protestant country. Her watch might have been a reminder of all she had lost.

The watch would also have had religious significance. In the sixteenth and seventeenth centuries, skulls regularly appeared in still lives and portraits, often accompanied by timepieces – sandglasses, clocks or early watches. The juxtaposition is fitting. Death and time of course go hand in hand: time is the unstoppable beat that counts down the hours of our life. Baudelaire once described it as 'the watchful deadly foe, the enemy who gnaws at our hearts'. Many skull watches were likewise inscribed with Latin quotations such as *tempus fugit* (time flies), *memento mori* (remember you are mortal), *carpe diem* (seize the day) or *incerta hora* (the hour of death is uncertain) to remind their owners that the afterlife, for which mortal existence was just a preparation, was waiting.

For Mary, the skull might also have represented the one domain over which she truly retained full control: her mind. Women at the time had very little freedom. Mary, who was just six days old when she succeeded to the Scottish throne, was a political pawn from birth. At six months old she was betrothed to Henry VIII's only son, the Protestant Prince Edward (future Edward VI), but the match was opposed by Scottish Catholics. At five years old, she was matched to Francis, then the Dauphin of France, and sent to be raised at the French court. When she returned home to Scotland as queen after Francis's death, she was to rule for just five years before her forced abdication in 1567. Even while under house arrest she was implicated in multiple plots by Catholic conspirators to usurp the throne of Queen Elizabeth I, until finally Elizabeth reluctantly signed her cousin's death warrant.

Imprisoned as she was for the last nineteen years of her life, Mary's skull watch might have been a reminder that, for all of that lost time, another, better, *eternal* life would begin after death. The once beautiful and vibrant queen's health suffered terribly in those two decades. She was plagued by bouts of sickness, including an agonising pain in her right side that stopped her from sleeping and intermittent discomfort in her right arm that made even writing hard. Her legs became so painful that, by the time of her execution, she was permanently lame. Later medical scholars have suggested that she suffered from a form of hereditary liver disease, or porphyria (known to have also afflicted one of her later descendants, George III), which might also account for her regular emotional breakdowns, attributed at the time to hysteria or madness. Her letters and the quotes she embroidered during her imprisonment make clear that she was ready to move on from her earthly life, in which the threat of execution hung over her like the sword of Damocles, and be reborn in heaven.

Mary was beheaded for treason at Fotheringhay Castle, Northamptonshire, on a cold February morning in 1584. She knew that in death she would herself become a symbol, and used the moment to secure her legacy for centuries to come. As she was led towards the block, her ladies-in-waiting assisted her in removing her jet-black satin dress, embroidered with black velvet, and embellished with jet buttons detailed with pearls. When they stripped back her outer garments, Mary was revealed to be wearing a deep-crimson petticoat, a red satin bodice and a pair of red sleeves: the colour of blood, and the Catholic colour of martyrdom.

I often picture Mary in her freezing cold chambers the night before her death, kneeling before her Bible in candlelight and holding her watch. The slowness of her timepiece (like all sixteenth-century watches) would have created a 'tick' closer in pace to the beating of her heart. I like to think that, as she prepared for her execution, her watch provided her with some comfort.

What was a watch in the sixteenth century? The truth is, it wasn't much of a timekeeper. Verge watches are temperamental. They had no devices to compensate for changes in temperature and didn't respond well to shocks or sudden movements. This made them prone to stopping and losing minutes, even hours. A watch was more of an exotic rarity, available only to the wealthiest. Very few watches were then in circulation and for much of the three centuries that followed their advent, they were exceptionally expensive. In a portrait of King Henry VIII by Hans Holbein the Younger from 1536, a curious locket hangs from a gold chain round the monarch's neck, looking an awful lot like a sixteenth-century watch. Further watches are listed in an inventory of the royal wardrobes of Queen Elizabeth I, including another skull watch and what might have been one of the first wrist-watches ever made. The list, drawn up in 1572, describes an 'armlet or skakell of golde, all over fairely garnished with rubys and dyamondes, having in the closing therof a clocke'. Unfortunately, this spectacular-sounding artefact hasn't survived.

The earliest watches typically fell into one of two styles that were fashionable to wear on a chain. One was more spherical, like Henlein's, and called a pomander after the cage-like balls containing fragrant incense that were swung in church. The other was housed in a flattened cylindrical case, known as a tambour from the French for 'drum', with hinged covers to the front and back. In the middle of the sixteenth century, technical advances allowed watchmakers to start exploring different case shapes, containing smaller mechanisms, as well as new techniques to decorate them. This was the beginning of form watches, so called because they mimicked the forms of other objects such as flower buds and animals. They also often doubled as devotional objects, taking the shape of crucifixes, Bibles and, of course, skulls.

Whenever I see form watches in museums or auction houses, I can't help but smile. They're such exquisite little objects. The perfect combination of intricate craftsmanship, design and engineering. If I had the freedom to create whatever I wanted and didn't need to worry about paying the bills, these are the things I would be making. They are as much *objets d'art* as watches, created collaboratively by masters of various crafts. They are often the products of enamellers, engravers, chasers, goldsmiths and lapidarists (cutters of precious stones) working together alongside conventional watchmakers, and are adorned with gold, silver, rubies, emeralds and diamonds. In these watches, function is still very much secondary to dazzling, unashamedly luxurious form. It's not hard to imagine how precious and emotive they would have been for their owners, and how they might have elicited a near-spiritual response to their cold stone or metal.

In 1912, workmen were demolishing dilapidated buildings on the corner of Cheapside in London when they saw something glittering under their feet. After carefully removing floorboards and soil, they unearthed 500 incredible pieces of Elizabethan and Jacobean jewellery – a discovery that was subsequently dubbed the 'Cheapside Hoard'. That story gives me goosebumps. One of the many magical properties of gold is that it doesn't tarnish with age or poor storage conditions. It's one of those materials that's instantly identifiable – too bright, too rich and warm in colour, and too heavy, to be anything else. Ornate vitreous enamel also keeps its colour like new, as do most gemstones. So despite spending nearly 300 years in the ground, this tangled heap of treasures would have gleamed almost as they did on the day they were buried.

At first the workmen decided not to declare their find to the property's owners, who happened to be the Worshipful Company of Goldsmiths.* Instead, they took the collection to jeweller George Fabian Lawrence, better known to London's navvies as 'Stony Jack'. They arrived with pockets, hats and handkerchiefs bulging with

---

* The Goldsmiths' Company is one of several wealthy London guilds that still own property all over the city.

jewels. Newspaper articles from the time describe them depositing 'many great lumps of caked earth' on the jeweller's floor, with one of the navvies exclaiming, 'We've struck a toy shop, I thinks guvnor!' Immediately spotting the significance of the collection, Lawrence set about securing the treasure for the London Museum.*

One of the most extraordinary items in the hoard was a form watch, its gilded movement set into a Colombian emerald the size of a cherry tomato. It is the only known surviving watch made with a solid emerald case. The precious green mineral has been cut in the shape of a hexagonal box to follow the prismatic shape of the raw crystal, so as to enhance its refraction of light and brilliance of colour. The dial, visible through its translucent stone lid, is of matching emerald-green enamel on a textured golden background embellished with a ring of Roman numerals. Unfortunately, the inner workings of the watch have not survived their subterranean adventure as well as their case has. Centuries of rust have fused the mechanism, tomb-like, inside its emerald crust.

What little we know about this watch has been deciphered from stereographic X-rays. The movement is typical of a finely made watch of its day, but unsigned, so we don't know who its maker was or where he was based. We know the emerald itself came from Muzo in Colombia – evidence of the international trade in luxury goods at that time – but we don't know who cut it. There were collectives of cutters capable of making a case like this in Seville, Lisbon, Geneva, and possibly London. Whoever made the case possessed astonishing skill. I trained briefly as a diamond grader, and know that to carve into a gemstone like this takes an incredible level of experience and ability. Lapidarists need an encyclopaedic knowledge of the properties of different stones. Every stone likes to be cut and finished in a different way – dictated by their molecular structure – and each one is like a fingerprint, with patterns of inclusions as unique as a snowflake. Lapidarists read the stones, studying their

* In 1976 the London Museum merged with the Guildhall Museum to become the Museum of London, which is where the Cheapside Hoard is now held.

structure in intimate detail, to figure out the shape of the final cut. Emerald is particularly tricky. It's part of the beryl family, and the shape of its molecules means it grows in hexagonal sticks, like a thick HB pencil. It's not as hard as ruby and sapphire, but chips easily. The larger and more precious the raw gemstone is, the greater the pressure on the lapidarist to get it right. One wrong move could chip a huge chunk out of the case or, worse, crack the whole thing in two – shattering its value at the same time.

We can be almost certain that this watch – in my view one of the finest jewels in existence – wasn't the work of one hand. Few form watches were: watchmakers in this era drew on the skills of a vast range of craftspeople to support and augment their work on the mechanism. To an extent Craig and I still do this today. Between us we have spent nearly four decades specialising in one craft, so we're painfully aware of what it takes to achieve mastery. We'll never be able to engrave, enamel or cut gemstones as well as a fellow artisan who has dedicated their life to the pursuit of their craft. If you want your work to be the very finest it can be, you have to collaborate.

Raw emerald naturally forms in hexagonal crystals.

Each time we design a watch, one of the first things we do is assemble an artisanal A-Team. Some of the people we work with are local (sometimes as local as the floor below us). Some are a few streets away. Some are from across Europe – true also of the watchmakers of the seventeenth century, although, unlike our horological predecessors, we met some of our contacts over Instagram. I have made a watch with a rock crystal case, sourcing the right stone from a dealer in London with the advice of a lapidarist in Birmingham who then cut it for me. I've collaborated with a dial enameller in Glasgow (the only commercial one left in the UK), sharing years of our research in order to help us achieve the perfect smooth finish.

I've commissioned a gun engraver in Germany to decorate our movements with ornate and minuscule acanthus-leaf scrollwork and text as small as the width of a grain of rice. Sharing your ideas with fellow specialists is a catalyst for innovation and creativity. For a job that can be so solitary, it's invigorating to be around others as dedicated to their crafts as you are to yours.

Acanthus leaf scroll engraving on a watch plate. The style has remained popular for hundreds of years – this is part of a design for a watch we made recently.

In the sixteenth century many of the best enamellers, engravers and goldsmiths worked in the small French city of Blois.* The Château de Blois was an official royal residence, and the local craftspeople catered to the royalty and nobility, who treated the city as a second home. This is where our skull watch was created. While these artisans undoubtedly earned their international reputation through their work, their renown was also assisted by their enforced movement around Europe during an era of violent religious persecution. It is powerfully ironic that Mary's watch, a potent religious talisman for a Catholic woman in an inhospitably Protestant country, was very likely made by a Protestant artisan in an equally hostile Catholic France.

---

* Blois had strong ties with the Medici family. Catherine de' Medici lived at the Château de Blois, dying there in 1589.

A great many exceptionally talented French craftspeople at that time, from goldsmiths and enamellers like those in Blois to watch-makers in Paris, were Huguenot – the popular name given to the French followers of John Calvin, founder of the protestant Calvinist Church. Catholic France under Catherine de' Medici and her sons was a brutal enemy of Protestantism. The nation's ruling Catholic aristocracy (many of whom were linked to Mary, Queen of Scots through blood or marriage, although Mary herself always practised and preached toleration between the Christian faiths) was responsi-ble for supporting atrocities that today would be regarded as ethnic cleansing and genocide.

Trouble began for the Huguenots in 1547, when Mary's father-in-law, King Henry II of France, decided to take direct action against the Protestant threat, condemning over 500 Calvinist followers to death for heresy. In 1562, the year after Mary arrived in Scotland, her uncle, the Catholic factional leader Francis, Duke of Guise, sent his soldiers to disperse a group of Huguenots holding a forbidden religious service within the town of Vassy. In the bloody resistance that followed, many of the 1,200 Huguenots in the congregation, including women and children, were slaughtered with swords and muskets. The massacre (news of which cannot have endeared Mary to her new Protestant Scottish subjects) tipped persecution of the Huguenots into an all-out war of religion. In August 1572, another Guise-led attack saw the loss of several thousand Protestant lives at the St Bartholomew's Day Massacre in Paris. As many as 10,000 were killed in similar riots in Bordeaux, Lyon and other French cities. Over the next two decades of persecution, thousands of Huguenots fled across Europe. They were refugees, arriving with little more than the clothes on their backs and the skills in their heads and hands.

The British Museum possesses a watch made by the French watchmaker David Bouguet, dating back to around 1650, that bears evidence of such foreign talent. It is a small rounded case, the shape of a classic pocket watch, measuring just over 4.5 centi-metres across, and covered with black enamel and a dense bed of brightly coloured enamelled flowers: blood-red roses, blue-and-yellow violas, variegated tulips and chequered red-and-white fritillaries connected by yellow gold and green scrolling vines. The front of the case, which covers the dial, is further embellished with ninety-two diamonds set in bands around the flowers. These old diamonds, each slightly irregular in shape, are cut in a style known as the Dutch rose. Crafted with far fewer facets than modern diamonds, whose many facets work together to reflect as much light back out of the stone as possible, they are duller but sparkle with a subtle grey fire, like droplets of water on the bonnet of a gloss black car.

More delights await inside. As you open the case, the cover above the dial reveals a painted country scene of a rambler walking with a cane, picked out in fine black linework on sky-blue enamel. The Roman numerals of the dial are in black on a white chapter ring that surrounds a full-colour miniature of two toga-clad figures in conversation amid an Elysian lakeside scene, while a flock of birds flies overhead. The decoration of the actual watch movement is, in contrast, quite restrained, though beautifully finished. It is very likely that the case of Bouguet's watch would have been created entirely by the artisans of the French city of Blois before being exported to London, where Bouguet, a newly arrived Huguenot watchmaker from France, created a movement to fit.

The Huguenot refugee community was close-knit. Its members shared knowledge, skills and crafts with their friends and passed them on to the next generation. David Bouguet's family is a perfect example of this process. Bouguet had arrived in England by 1622, and in 1628 he was admitted to the London Blacksmiths' Company. Two of his sons, David and Solomon, also became watchmakers. Another of his sons, Hector, apprenticed a Huguenot diamond cutter, Isaac Mebert/Maubert, who ended up marrying Bouguet's daughter Marie. Isaac's brother Nicholas married Bouguet's other daughter Suzanne, and another of his daughters, Marthe, married a jeweller (Isaac Romieu). I wouldn't be surprised if the diamonds on Bouguet's black watch were cut in his son-in-law's workshop.

Londoners often referred to Huguenots as 'liberties' or 'strangers', and didn't always make them feel welcome. In 1622, English watchmakers were so worried about the new arrivals that they petitioned James I to prevent them trading in the city and to establish a specialist livery company for watch- and clockmakers. The Worshipful Company of Clockmakers was duly founded in 1631, though Huguenots weren't actually excluded from it. In 1678 the Goldsmiths' Company, complaining that Huguenots were undercutting English workers and damaging their trade, attempted to prevent foreign Protestant craftspeople from working in certain places and accessing the seven-year apprenticeship that allowed them to become a member of the Company. As it happened, many of the anti-French campaigners employed descendants of the Huguenots and it became common for British craftsmen to send their sons to study with a Huguenot master or take on a Huguenot apprentice themselves. Bouguet's life must have been one of highs and lows: one minute he was working for wealthy patrons who admired, respected and valued his work; the next, he was being called a 'French dog' (or worse) in the street.

The rise of Reformed Protestantism fundamentally changed the nature of the watch. The first things to go were the signs of ostentatious wealth. Where Catholic watchmakers glorified God with elaborate decoration, the Reformed worldview saw decoration as a distraction from His true glory. John Calvin even banned his followers from wearing jewellery. This, ironically, prompted many jewellers to turn to watchmaking, leading to the rise in fine goldsmithing, enamelling and stone setting in Swiss watch cases.*

In Protestant England, the Civil Wars of the 1640s and Oliver

---

* Genevan watchmakers who wanted to make more elaborate pieces found markets for their work abroad. The small silver watch in the form of a pocket-sized lion made by Jean-Baptiste Duboule in around 1635 was very likely destined for Constantinople in the Ottoman Empire.

Cromwell's reign in the 1650s saw Puritans try to 'purify' the Church of England of every last vestige of Roman Catholicism, of which they felt too much remained under the rule of Charles I. Flamboyant dress was criticised for reeking of 'papery and develyshnes' and was seen as a symbol of pride and an incitement to lust. Everything from beard curling, imported fragrances and ruffs to cosmetics, tight doublets and presumptuously large codpieces came under attack. Even the powdered wigs that dominated fashions under Charles I were given the chop. Puritans dressed modestly, wearing sober colours, plain linen cuffs and collars (sometimes from home-woven cloth without any trimmings or buttons at all) and straight, simple hair.

A typical silver Puritan watch, plain and lacking in ornate decoration.
They were a total contrast to the earlier form watches.

Watches were already such invaluable tools that even strict Puritans didn't want to give them up (Cromwell himself seems to have owned one). But they too were simplified significantly. Puritan watches were relatively small, measuring around 3 centimetres across and up to 5 centimetres in length, and often oval in shape. They were devoid of decoration or embellishment, without gemstones or fritillaries. Their cases were usually crafted from silver – gold would have been too ostentatious* – and completely plain, their smooth surfaces not unlike the gentle sheen of a weather-beaten pebble on a beach. Their dials, hidden inside the front cover, were also plain, with the exception of their purely functional chapter ring, indicating the hours of the day by a single hand.

In this new pared-back form, the watch proclaimed a very different understanding of time from the skull watch that started this chapter. The Protestants thought of time as a gift from God; to waste it was a sin. They believed that to prosper in the next life, you needed to use your time well in this one. Puritan values emphasised responsibility, self-control, hard work and efficiency. In the Puritan day there was no such thing as spare time, only time that should in fact be spent in God's service. It was even argued that 'spending time in leisure activities' was 'a form of theft, a defrauding of the master'.

In 1673, the influential Puritan church leader Richard Baxter published *A Christian Directory* in order to guide the faithful as to how a good Christian should manage his or her time:

> Time being man's opportunity for all those works for which he liveth, and which his Creator doth expect from him, and on which his endless life dependeth, the Redeeming or well improving of it, must needs be of most high importance to him: And therefore it is well made by holy *Paul* the great mark to distinguish the *Wise* from fools.

Time management, it would seem, was godliness. 'One of the greatest Time-wasting sins is idleness, or sloth,' he writes, condemning:

> He [who] spendeth his Time in fruitless wishes: He lyeth in bed, or sitteth idly, and *wisheth,* Would this were labouring: He feasteth his flesh, and *wisheth* that this were fasting: He followeth his sports and pleasures, and *wisheth* that this were prayer, and a mortified life. He lets his heart run after lust, or pride, or Covetousness, and wisheth that this were heavenly mindedness.

---

* Although, remarkably, exceptionally rare examples in gold do exist.

> . . . See that ye walk circumspectly, says the Apostle . . . redeem-
> ing the time; saving all the time you can for the best purposes; buying
> up every fleeting moment out of the hands of sin and Satan, out of
> the hands of sloth, ease, pleasure, worldly business.

I've spent a large portion of my life feeling guilty: for not working hard enough, for oversleeping. Even on holiday I struggle to relax, because of the guilt of not working. I doubt that my ancient ancestors, living in caves carving star charts, felt the same sharp pang of shame when they took a moment to relax. Time guilt is rooted in social conditioning. Having a lie-in on a Sunday morning isn't going to make much of a difference to anything. Stealing a precious extra few moments in a cuddle with your child, or pausing in the garden to enjoy the sunshine on your face, might not get the work finished, or the washing-up done – but the guilt many of us feel about such activities is often completely disproportionate to their impact. Something in our cultural history has taught us to feel bad for not working. Reading Baxter on the mortal perils of enjoying a moment's pleasure, I can't help but suspect that just as the rituals of Catholicism continue to stir me, so sixteenth-century Puritanism runs through my non-believing bloodstream. Though this extreme interpretation of Christianity has been marginalised for more than 300 years, its teachings continue to infiltrate our experience of time. Puritanism spelled the beginning of the end of the 'work–life balance'.

Cromwell's Puritanical Commonwealth was short-lived. After his death in 1658, and a brief rule under his son Richard that lasted less than a year, a king returned to the English throne. Decadent timepieces predictably followed. Charles II was a great admirer of the artistry of horologists. He kept at least seven clocks in his bedroom – their poorly synchronised chimes drove his assistants to distraction – and another in the antechamber that also recorded wind direction. As his reign progressed and watchmaking flourished, he often insisted on being the first witness to the latest horological inventions.

In France, the Edict of Nantes in 1598, signed on the ascension of Henry IV, gave the Huguenots some years of relative peace. By the 1680s, however, although Louis XIV had pledged to support the Edict, a renewed campaign to purge France of Protestantism was under way. Through forced conversions, propaganda, the separation of Huguenot children from their families and the demolition of Protestant temples, life for Huguenots became increasingly difficult once again. Finally, in 1685, the 'perpetual and irrevocable' Edict collapsed. What followed was another Huguenot exodus. In the years that followed the Revocation, between 200,000 and 250,000 Huguenots escaped, while as many as 700,000 renounced their faith and converted to Catholicism. While the majority of refugees headed to the Dutch Republic, the second-most popular destination was Britain, where between 50,000 and 60,000 refugees are estimated to have fled. Switzerland also provided shelter for a great number of Huguenot settlers. These Huguenot migrants played a central role in the development of the watch industry both in the UK and in Switzerland, and their impact still resonates to this day. Ruled over by a clock-loving sovereign, and now enriched, by an influx of new talent, London was set for a golden age of watchmaking.

4

The Golden Age

*Out of the right fob hung a great silver chain, with a wonderful kind of
engine at the bottom . . . He called it his oracle, and said it pointed out the
time for every action of his life.*

Gulliver's Travels, Jonathan Swift, 1726

As a trainee watchmaker, the first thing you must do is make
your tools. It makes sense that, before you're allowed anywhere
near the incredibly fragile mechanics of a watch movement, you
start on something robust that you'll need in the long run. My first
project, on my three-year course run by the British Horological
Institute, was making a 'centre cutter/scriber'. This was a pencil-
like rod of steel with two different functions. I had to finish one end
as a flat-headed screwdriver, only razor sharp (the scriber), and at
the other I had to create three facets that come together at their tip
(the cutter). This we would use to 'key' a metal surface, cutting a
tiny recess to create a location point for a drill bit to bite into.
Watchmakers need to drill a lot of holes.

We used our first tool to help make our next – a movement
holder. Like a tiny vice, it is the bit of kit you use to hold a watch
movement while you work on it. We had to file the holder by hand
according to a precise technical drawing and our finished piece had
to be accurate to those dimensions within three-tenths of a millimetre. This tiny amount of leeway for accuracy is called tolerance.
It felt incredibly precise at the time but this was just a

beginner's-level introduction to the microscopic world of watch-making. These days, we sometimes work within a tolerance of microns – thousandths of a millimetre.

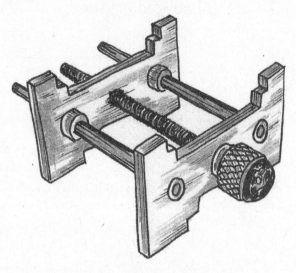

A brass movement holder designed to hold a
watch mechanism while I work on it.

In this way we inched through our training, only touching an actual pocket watch movement (which I worked on in my new movement holder) for the first time towards the very end of our first year. Several of my classmates had dropped out by this stage, fed up with only working with files and bits of metal for the best part of a year. For me, making the movement holder was like a very controlled form of jewellery and silversmithing, so I loved it. But even then I couldn't resist adding a decorative flourish to this most basic of tools. I decorated it with a scratch-brush grained finish I'd learned as a jeweller (rather than straight-grained, which is the 'proper' horo-logical way). I gilded it, and commissioned a lapis lazuli to be cut by our local lapidarist that I then set in the winding button you turn to open and close the jaws. Project number two and I was already doing things I hadn't been told to.

By the second year we were starting to make watch parts – over-size ones at first, using drawings from the syllabus as reference, as we refined our skills to work on components for the real watches on our bench that were ever smaller and more intricate. We tackled pocket watches at first, before slowly working our way down to smaller gent's-size and finally tiny lady's-size mechanisms. We learned to service the most basic of complications, like automatic winding work or date and calendar mechanisms, before gradually progressing to the world of chronographs. The syllabus also required we demonstrate our competence with the verge, the cylinder, and the English and Swiss lever escapements. Although all but the last are redundant in modern watchmaking, for a restorer they will never die out. After the foundation course, graduates could get an entry-level job in a workshop and from there it would take several more years of hard graft and luck to work your way up to being a master watchmaker. It was an education in dexterity, attention to detail and, most of all, patience.*

Things were by no means easier in the eighteenth century. A watchmaking apprenticeship, a legal requirement to practise as a watchmaker in the City of London, was seven years of almost monastic intensity (apprentices were not allowed to marry while training). This would then be followed by perhaps two or three years as a journeyman watchmaker honing your skills, until the completion of a 'master-piece' – a complete watch from scratch – earned you the title of watchmaker.

Highly skilled and inventive, the best watchmakers were in fierce demand and had begun to enjoy high status and renown. This was the golden age of English watchmaking, when Europe's most brilliant watchmakers bounced ideas off one another and vied to advance the accuracy and complexity of the watch. Many of the famous watch- and clockmakers of the day would have been known to each other through the Worshipful Company of Clockmakers, a livery company established in 1631. In archival documents from this

---

* I didn't know then that I would be one of the last to receive this training. A few years later the course was axed in favour of a more theory-heavy academic BA in horology.

era, I've found the names of the greats all listed together as signatories of the same letters from the Company, like a roll-call of the celebrated names in horology: Thomas Tompion, regarded as the 'Father of English Clockmaking', who worked with Robert Hooke to create some of the very first watches with balance springs; George Graham, Tompion's pupil and successor (he married Tompion's niece, Elizabeth Tompion), who, when he wasn't busy making scientific instruments for Edmond Halley, found time to invent the orrery and made considerable improvements to pendulum design; Daniel Quare, master of the repeating watch, and Thomas Mudge, once George Graham's apprentice, George III's Royal Watchmaker, whose lever escapement was quietly revolutionary. These are the celebrities of the watchmaking world – a workshop employing them all would be the equivalent of a Premier League fantasy football team. In a century that saw the invention of the piano, the steam engine, the hot-air balloon, the spinning jenny and the steamship, the watch kept pace, proving vital to solving some of the most pressing scientific questions of the age.

By the start of the eighteenth century timekeepers were physically, if not always financially, accessible, and familiar to all. The vast majority of English parishes now possessed a public clock in their church tower and clocks had begun to appear in inns, schools, post offices and almshouses. By the end of the century there would be clocks hung in every pub and tavern throughout the British Isles. As you walked the streets of cities such as London or Bristol a clock was never far from sight or hearing. Clocks were increasingly found in domestic homes – even servants, who still could not afford to own a clock themselves, would have been used to seeing them. For those who owned domestic clocks, the most popular location for them was the kitchen – one of the few rooms that every home has, regardless of your wealth and status.

Time didn't only enter public consciousness but became the subject of intense philosophical debate. While Isaac Newton believed that time was '[a]bsolute, true, and mathematical', others such as David Hume and John Locke argued that it was relative – that time essentially depended on the people perceiving it (these ideas were famously taken further in the twentieth century by Einstein's theory of relativity). In the same period, the writer Laurence Sterne gleefully played with time in his masterpiece, *The Life and Opinions of Tristram Shandy, Gentleman*, crafting a narrative in which time contracts, expands, and goes backwards as well as forwards.

Watches, from the second half of the seventeenth century onwards, had occupied an increasingly important (if not especially useful) role in the personal lives of the wealthy. Samuel Pepys, picking up his new watch from Briggs the scrivener ('and a very fine watch it is') in May 1665, was no less enamoured, distracted and ruled by his new watch than any of us might be with a new smartphone. 'So home and late at my office . . .' he wrote, as he so often did,

> But, Lord! to see how much of my old folly and childishnesse hangs upon me still that I cannot forbear carrying my watch in my hand in the coach all this afternoon, and seeing what o'clock it is one hundred times; and am apt to think with myself, how could I be so long without one; though I remember since, I had one, and found it a trouble, and resolved to carry one no more about me while I lived.

Two months later his watch was already back at the menders. Accuracy, for the watch, was still a work in progress.

Luckily for Pepys, the late 1600s saw two remarkable leaps forward in horological progress that kick-started a century obsessed with time. In 1657 the Dutch mathematician Christiaan Huygens invented the pendulum clock, successfully applying Galileo's theory of isochronism of 1637. The theory is that a pendulum will take the same duration to swing regardless of how big that swing is. This consistent swing triggered the equally consistent hold and release of the escapement, forging the way for the most accurate kind of clock that had ever been invented. The long pendulum swinging underneath lent itself to the longcase design, or 'grandfather' clocks, which became increasingly popular in the decades that followed. In 1675 polymath and scientist Robert Hooke invented the metal hairspring, or balance spring as it is also known, as revolutionary for the watch as the pendulum had been for the clock. A flat steel spiral of very fine wire, the metal hairspring* was designed according to the principle of Hooke's eponymous law of elasticity – '*ut tensio, sic vis*', meaning 'as the extension, so the force'. Force exerted on a spring causes an equal return of force by the spring. If I tighten a coiled spring beyond its resting position and then release it, the spring will fling itself outwards. But as it flings out, the force will cause it to fling out a bit too far and then want to coil back again in order to return to the nice comfy spiral in which it was first formed. This action causes the spring to 'breathe' in and out as it over- and under-coils rhythmically with the oscillation of the balance wheel, triggering a consistent release of the escape wheel teeth. The effect on accuracy was dramatic. It was this invention that, for the first time, made it worthwhile adding a minute hand to a watch – a milestone in the history of mechanical watchmaking. Metal hairsprings were quickly welcomed by an increasingly watch-loving public and it was not unusual for people to have metal hairsprings retrofitted to their older, pre-hairspring watches, which had little or no regulation. Pepys was soon timing his walks between Woolwich and Greenwich to the minute.

The City of London was the ticking heart of the watchmaking world at this time. In 1665, the Great Plague had devasted the population of London (in eighteen months it killed almost 100,000 people, nearly a quarter of the city's inhabitants), but our Huguenot artisans, fleeing the Revocation of the Edict of Nantes in 1685, had swelled its numbers. By the eighteenth century, watchmaking workshops in England were usually small collectives of journeymen and apprentices, all led by a master and supplemented by the work of

---

* Metal hairsprings replaced the literal hog's-hair bristle that preceded them.

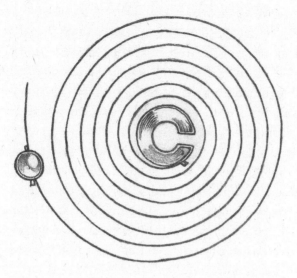

Early hairsprings were formed as a flat spiral. The collet at
the centre is used to secure it to the balance staff.

several other local craft workshops such as goldsmiths, engravers,
chainmakers and springmakers. When I take apart a watch from this
era, I can often count as many as four or five, or even more, different
makers' marks and other signatures, concealed throughout the
entirety of the movement, inside and out. In this way I can mentally
piece together the creative colonies that existed in watchmaking
centres like Clerkenwell in London, with watches and watch parts
being shifted around multiple workshops all within a few streets'
radius of each other. When I study old maps, looking up the
addresses of registered makers, I can't help but notice there's often a
tavern or inn in the middle of it all, and I like to imagine these
craftspeople huddled together over jugs of ale in a smoky tavern
discussing business and sharing ideas. There are still echoes of this in
Birmingham's Jewellery Quarter, where I work today. Although
local boys no longer run down the road to the assay office with
wheelbarrows piled with gold jewellery to be hallmarked, we makers
still look out for each other. We all know each other's business and
sometimes like to meet for a pint of real ale.

Watch tools, components and later even whole movements (though not yet cased and ready for retail) were often made in workshops in Lancashire, particularly in Prescot, 8 miles east of Liverpool, where ample coal supplies, a long-standing tradition of metalworking and a good transport link to London fostered a cottage industry of suppliers, but the most advanced branches of the trade were concentrated in London.

Outside London, watch- and toolmakers rarely completed an apprenticeship, but within the city, the tightly controlled apprentice system created valuable opportunities for those trainee watchmakers who were lucky enough to find a position. Master watchmakers and their formal apprentices would move around different cities, widening their network of potential patrons. Their watches were charged at prices based as much on their social status as the quality of their work, so watchmakers (and their apprentices) from wealthier backgrounds were set up to be more successful from the start. It was a bit like the watchmaking version of public school – it was not just the education but who you met that conferred the advantages.

Fifteen-year-old Thomas Mudge, the son of a headmaster in Devon, arrived to take up his apprenticeship with the renowned clock- and watchmaker George Graham in the spring of 1730. On gaining his freedom of the Clockmakers' Company in 1738, he took lodgings and spent the first part of his career in the shadows, making exceptionally complicated watches on commission to other watchmakers, signing his works with their names – a standard practice at the time. He might have carried on in this way indefinitely were it not for a watch he had made for another celebrated maker of the day, John Ellicott. The watch, which could display the equation of time (the difference between the true solar day, which changes according to the position of the sun, and the mean or average solar day) with a range of additional calendar indications, was sold to King Ferdinand of Spain. But, so the story goes, someone in the royal court dropped the watch on the floor, damaging it so badly that it had to be sent back to Ellicott, who was unable to fix it. He was forced to send it back to its actual maker, Mudge, who undertook the necessary repairs. When King Ferdinand found out, he insisted on commissioning Mudge directly in the future. The king's patronage lifted Mudge from obscurity, encouraging him to make the leap to setting up his own workshop at 'the Dial and one crown' on Fleet Street in 1748.

Mudge became renowned for the mechanical innovation of his watches. He made King Ferdinand of Spain a grande sonnerie watch set into the top of a walking cane. Grande sonneries are considered to be one of the finest complications as their mechanism chimes both the hour and the quarter-hour, but can also chime out the time on demand if the owner wants to hear it in between. Mudge was also famous as the first watchmaker to integrate a perpetual calendar – so called because it compensates for variations in the duration of months and years to 'perpetually' show the correct day and date – into a watch. Thomas Tompion and George Graham had applied the perpetual calendar to a clock as early as 1695, but the process of shrinking the mechanism down into something small enough to fit into a pocket watch is credited to Mudge.* Just as powerful patronage had supported the development of the first mechanical clocks in the days of Su Song, so the technical advances of these ever-more intricate watches were bankrolled by the wealthy who desired them. It was the

---

* Perpetual calendars are possibly one of the most discreet, yet complex, horological indications invented. They take something we might take for granted on a modern watch, a date display, and turn it into a date that is displayed with accuracy in (almost) perpetuity. To do this they employ a series of gears that can count not only the days and months, but years and leap years. They memorise the number of days in each month for us in the teeth of their wheels – the slowest of which turns once every four years. Looking at one of Mudge's surviving pocket perpetual calendars, dating from around 1762, the care he took to ensure that the dial was functional and easy to read is evident. The date is marked by a gold marker set above the twelve-o'clock position, which sits against a rotating disc of dates on the very outer rim of the dial. In the centre of the dial the moon phase is shown above two crescent apertures that look onto dials showing the day and month. February has its own dial within a dial indicating whether it is a leap year. And yet, despite the level of detail the watch is carrying, nothing is fussy.

equivalent of investing in a tech start-up. Your investment helps the business innovate, and that innovation boosts the value of your shares. Only instead of shares you receive a very lovely watch you can wear as it (ideally) increases in value along with the renown of its maker.

Mudge's correspondence with another of his patrons, Count von Brühl, a Polish-Saxon statesman reputed to have owned the largest watch collection in Europe, gives us an insight into the mutual rewards of the patron–craftsman relationship. The two were in regular contact throughout the making process. In his letters to von Brühl, Mudge goes into a surprising level of technical detail, discussing engineering and material principles, temperature coefficients and other technical challenges he faced. For many horological patrons, commissioning a watch was more than just the purchase of a beautiful thing; it was a dynamic relationship in which the patron invested as much in the process as he or she did in the finished product. The patron often had a genuine desire to fully understand their watch's workings and feel part of its creation.

In our workshop, the involvement of collectors and patrons is still a key aspect of the making process. Working bespoke not only gives us the opportunity to meet a specific brief but also to iron out any requests and alterations while a watch is being made. We've remade and adjusted the size of winding crowns to suit the ergonomics of our clients' wrists, adapted them if they have arthritis and struggle with winding a watch, and adjusted them depending on whether they wear their watch on the right or left. We've altered colourways and proportions to refine legibility, we've repeatedly modelled and remodelled cases for clients who give us feedback on how they feel compared to their other watches. The client's involvement in the many mini-decisions that are part of the creative process is immensely useful for us, but it also makes space for them to become part of our workshop. They become as much a part of the finished product as our own hands have been.

The patron or collector who commissions a watch also brings a wider perspective to us as makers. One of the challenges of being a watchmaker is that we rarely earn enough to be able to afford the objects we make. The result is that we don't tend to have significant collections ourselves, certainly not at the level our patrons do. It's why I find our clients' input so valuable: they're the ones out there in the real world looking for pieces; they know what they choose and why, what pieces are actually like to own and use on a day-to-day basis.

In Mudge's era, a patron was also keen to be associated with the very latest scientific developments. Mudge's association with von Brühl led to a purchase by King George III, who in 1770 commissioned Mudge to make a watch for his wife, Queen Charlotte, featuring the earliest known example of Mudge's most groundbreaking invention, the detached lever escapement. The king, like von Brühl, took an active interest in the watches and clocks he commissioned, and was something of an amateur horologist himself. There are manuscripts in the Royal Collection written in the king's hand in which he describes the process for assembling and dissembling watches. Queen Charlotte too was an avid watch collector and something of a magpie. Her friend, the diarist Caroline Lybbe Powys, recalled seeing 'twenty-five watches, all highly adorn'd with jewels' in a case beside the queen's bed at Buckingham House in 1767. I can't help but read into the fact that she kept her collection in such a close and personal proximity, falling asleep by them each night and waking up to them every morning. Queen Charlotte clearly loved her watches.

The 'Queen's watch', as Mudge came to call his watch for Queen Charlotte, would certainly have won King George brownie points. Mudge later described it as 'the most perfect watch that can be worn in the pocket, that was ever made'. From a watchmaker's perspective, it is Mudge's detached lever escapement that is the true star.*

One of the great enemies of accuracy in the watch is friction, because it upsets the precision of the escapement. In the verge

* Mudge first developed the lever escapement in 1754, but it appeared for the first time in his Queen's watch.

escapement, the oscillating balance wheel was in near-constant engagement with the train wheels, creating varying friction. Mudge's great innovation was to invent an escapement that was 'detached' from the balance wheel. His lever worked by being knocked back and forth by a pin fixed to the underside of the oscillating balance wheel. As the wheel, and pin under it, swung back and forth they'd flick one end of the pivoted lever back and forth at the same time. At the other end of the lever, a pair of pallets would catch and release a tooth of the escape wheel with each 'tick'. This meant that the oscillating balance was exposed to friction for only the very briefest moment as the lever was flicked across.

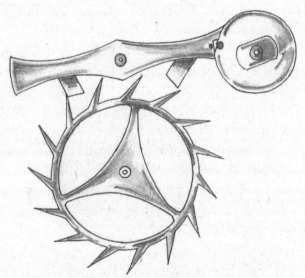

The English lever escapement – the commercial
adaptation of Thomas Mudge's detached lever. The design
was refined in the nineteenth century to form the Swiss
lever escapement (spot the national rivalry), which is still
used in almost all mechanical watches to this day.

Mudge himself was modest about the potential of this invention, if only because of its extreme technical intricacy. In a letter to von Brühl, he declared the level escapement:

requires a delicacy in the execution that you will find very few artists equal to, and fewer still that will give themselves the trouble to arrive at; which takes much from its merits. And as to the honour of the invention, I must confess I am not at all solicitous about it; whoever would rob me of it does me honour.

I'm not surprised that they gave Mudge a headache. Making lever escapements requires a high level of accuracy, and is challenging to pull off even today, when we have the benefit of modern engineering equipment. Yet we still do honour him by 'robbing' it. To this day, virtually every mechanical watch made in the world uses his invention.

This kind of high-level technical innovation meant that watchmakers began to play a crucial role in the development of other industries. Horologist Samuel Watson helped physician Sir John Floyer (1649–1734) design, make and sell the first pulse watches to aid physicians in counting their patients' pulses. Thomas Mudge created a watch for John Smeaton (1724-1792) with temperature compensation, to even out the expansion and contraction of metals as they move between hot and cold temperatures. Smeaton, the world's first self-proclaimed 'civil engineer', is famous for his work developing new forms of cement, which he used to improve the way Britain constructed its lighthouses. He used his Mudge watch to assist him in his survey work. (A device to compensate for changes in temperature also appears in Mudge's watch for Queen Charlotte). In 1777, watchmaker John Wyke supplied all the wheels, pinions and framework to make the first pedometer, designed by Matthew Boulton, a kind of antiquarian Fitbit capable of counting the wearer's steps. Watch- and clockmakers were recruited to help refine and maintain factory equipment. According to a 1798 account from a factory in Carlisle, '. . . the cotton and woollen manufactories are entirely indebted for the state of perfection to which the machinery used therein is now brought to the clock and watch makers, great numbers of whom have, for several years past . . . been employed in inventing and constructing as well as superintending such machinery . . .' But the most famous horological challenge of the eighteenth century, an age in which the British navy expanded exponentially, was that of the 'quest for longitude' – the essential navigational coordinate that until now had relied on the stars, dead reckoning, a sand-timer and guesswork.

The starter gun that announced the beginning of the British quest for longitude was fired early on in the century, triggered by one of the worst maritime disasters in British naval history. On a foggy night in 1707 four Royal Navy warships under the command of Sir Cloudesley Shovell were wrecked on rocks off the coast of the Isles of Scilly. The warships, on their return from Gibraltar after besieging the port of Toulon in France, sank, and most of their crew drowned, their bodies washing up onto the surrounding coast for days afterwards. As many as 2,000 men were lost. The cause of the tragedy was deemed to be a fatal combination of poor visibility and a mistake in their course caused by plotting their longitude incorrectly, meaning that they were completely unaware of the approaching hazard.

The ships had struggled to find their exact location or 'navigational position'. On land, the process of establishing a navigational position is relatively straightforward, using landmarks as reference. At sea navigators were more, well, *at sea*. Latitude – how far north or south of the equator you are – could be determined from the position of the sun in the sky. But to calculate the east–west position, known as longitude – how far you've travelled through imaginary lines that run from pole to pole – a navigator needed to be able to calculate speed and course from a given position at a given time (usually the home port and the time of departure). Wind, currents and tides all affected calculations, while motion and temperature could affect the accuracy of any timekeeper, with potentially fatal results.

The origins of pinpointing longitude date back thousands of years. Some of the stars we still use as navigational markers today are mentioned in Homer's *Odyssey*, in which the goddess Calypso tells the hero, Odysseus, how to steer his ship on a steady course to Phaeacia by keeping the stars of the Great Bear on his left. The credit for devising longitude at sea most likely originates with the Polynesians, who for thousands of years had been masters of oceanic navigation. Tupaia, a Tahitian Polynesian navigator recruited by Captain Cook in 1769 during his expedition onboard HMS *Endeavour* to map Terra Australis Incognita, astonished the crew with his almost instinctive ability to know precisely where they were using the stars and dead reckoning. He was also able to draw a now-famous map of vast tracts of the Pacific from memory – roughly the area of Europe including European Russia – including the names of seventy-four islands, and give detailed accounts of the complex Pacific wind system. What I find most striking about this is that the means by which Polynesian navigators like Tupaia found their way around the Pacific Ocean were natural, just like those we first used to discover and measure time. By the eighteenth century, however, Europeans had become so disconnected from the natural world surrounding them that they needed a machine to help them.

The Scilly naval tragedy proved a catalyst. In 1714 the Longitude Act announced a prize of £20,000, about £1.5 million in modern money, to the person who could solve the longitude problem. It laid down a challenge for Britain's greatest minds across the fields of science, engineering and mathematics. The committee sought advice from Isaac Newton, by then the mature age of seventy-two, and his friend Edmond Halley, whose travels mapping the stars made him an obvious choice. When Newton presented his remarks on the task ahead to the committee, he listed the methods that were currently in existence, albeit 'difficult to execute'. One method, he said, 'is by a Watch to keep time exactly. But, by reason of the motion of the Ship, the Variation of Heat and Cold, Wet and Dry, and the Difference of Gravity in different Latitudes, such a watch hath not yet been made.' And nor, in his opinion, was it likely to ever be.

Newton and his contemporaries felt sure the solution lay in astronomy – perhaps in studying the eclipses in Jupiter's satellites, or in predicting the disappearances of stars behind our moon, or by observations of lunar and solar eclipses. There was also the possibility of a lunar distance method, where longitude could be calculated by measuring the distance between the moon and the sun by day, or the moon and navigational stars by night. This feedback was used to dictate the terms of the prize, which awarded a first, second and third place to would-be applicants from any scientific or artistic discipline for inventions judged purely on their degree of accuracy when tested 'over the ocean, from Great Britain to any such Port in the West Indies as those Commissioners Choose . . .' (In other words, the Caribbean–UK leg of the trans-atlantic slave trade triangle . . .).

With the most renowned minds of the day called to action, no one expected the answer to come from an unapprenticed clock-maker from Yorkshire, or to arrive in the shape of a watch.

At first glance, H4 looks like a typical pocket watch of the era, although with an overall diameter of 16.5 centimetres, you would struggle to find a pocket large enough to fit it. Aesthetically too, it looks similar, with a plain polished silver case and a dial made from white enamel and marked with black numerals bordered by decorative scrolls of acanthus leaves detailed in black linework. But this was no ordinary watch. Weighing nearly 1.5 kilograms (close to four tins of baked beans), it contains an extraordinary movement.

Completed in 1759, H4 is the fourth of John Harrison's five experimental marine timekeepers, made to satisfy the demands of the Board of Longitude. Its predecessors, H1, 2 and 3, had been large, unwieldy clocks, though of sufficient promise and technical brilliance to warrant the invaluable interest and support of George Graham, Mudge's old master. Even with Graham's backing,

Harrison's labour was a dogged, solitary effort. He was his own harshest critic and, discerning a fault in H2, would not allow it to be tested. Twenty years elapsed between H2 and H3, as Harrison struggled on, beset by technical difficulties. With H3, lighter than its predecessors and comparatively compact at 2 feet high and 1 foot wide, it seemed as though he had pushed the size of a sea clock to its limits.

Thus the arrival of the comparatively diminutive H4 just a year after H3, and in the form of a watch, came as something of a surprise. Harrison's H4 mechanism involved unprecedented engineering. While he used mechanisms that were already in watches at the time, such as the verge escapement, he refined them to a level rarely seen even now. To reduce friction and improve durability, the steel flags that make the entrance and exit pallets of the verge were made from diamonds. The round balance wheel is huge compared to a normal eighteenth-century watch, but this technical adaptation makes it less susceptible to variations as the watch moves with the motions of the ship. A longer mainspring gave the H4 a running time of thirty hours from full wind.

H4 was first put to the test in 1761 when HMS *Deptford* set sail from Portsmouth, headed to Kingston, Jamaica. Harrison sent his timekeeper to travel on board the ship, along with his son William as its minder. H4 was quick to make an impression, helping to correctly calculate the time of their arrival at the port of Madeira on the outward journey, earlier than the crew's own predictions. The captain was so impressed, it's said he offered to buy Harrison's next timekeeper on the spot. Over the journey, which lasted eighty-one days and five hours, H4 lost a mere 3 minutes and 36.5 seconds. It was considered to have met the Board's exacting requirements – subject to a second test. While Harrison wrangled with the Board to meet conditions that seemed to become ever stricter, H4 became a template for its successor, H5.

Today, Harrison's first marine chronometers lie in state at the Royal Observatory in Greenwich. In their decoration, H4 and H5 are far plainer than personal pocket watches of their day. Harrison's

H4 still bears some of the characteristic acanthus-leaf scroll engraving and pierced embellishment that was popular at the time, but by the start of the nineteenth century this ornamentation had been lost from the production of chronometers.

It seems that the more accurate and functional the watch, the less it needed ornate decoration to justify its existence. Nevertheless, these first scientific instrument watches are exquisite, just in a different way. Their beauty is in their functionality. I have to remind myself that they were made without any of the modern technology I have access to today, and yet they are still more accurate than some run-of-the-mill mechanical watches currently on the market. And, as I handle them, I am awed by the care that has been taken in their finish. I often notice a technique once known as chamfering, and now commonly referred to in the Swiss-French as *anglage*, where sharp corners are carefully filed to a 45-degree angle around the plates, bridges and even tiny spokes of every wheel in the train to make them appear lighter and more refined. These chamfers are brightly polished, in contrast to the grained or frosted flat surface next to them, so they glisten as you move the mechanism in the light. Although they can now be done perfectly by machine, you can tell when they've been done by hand as they catch the light differently. I'm also excited to see what we watchmakers call *black polishing* on some of these early chronometers.

Black polishing is typically only done on very hard metals like steel. When the surface has been polished to a mirror-like state of perfection, without a single scratch or mark, it appears as black as onyx when it catches a shadow. Black polishing was, and still is, done by hand and is immensely time-consuming. Polishing parts is useful to reduce friction, but polishing to this level is a demonstration of sheer skill by the maker. Even as precision and functionality became the singular purpose of these timepieces, it delights me to see how their makers found ways to include their own personality and identity in the finishing of their work.

For me, chronometers are one of the purest examples of how the watch is so much more than a piece of scientific equipment. It is a piece of scientific equipment crafted by human hands, sometimes over a period of years. Their finishing is as idiosyncratic as a signature, a mark of pride that reveals the maker's personal investment in the work, one that goes beyond pure function and makes it a work of art.

As a student watchmaker, I was raised with the belief that John Harrison was a horological hero whose invention had saved incalculable lives at sea. In many ways that was true; Harrison *was* a brilliant inventor and his timepieces had a dramatic impact on navigation. But the H4 was not altogether the solution it is fabled to be.

In 1831 HMS *Beagle* carried twenty-two watches and chronometers, as well as a young and enthusiastic university graduate named Charles Darwin, onboard her second voyage to survey the coasts of the southernmost parts of South America. By the time she returned home in October 1836, only half of the timekeepers were still in good working order.

One of the problems early chronometers faced was accumulated inaccuracy over very long distances. If the inaccuracy was consistent – say, if the navigator knew the chronometer was gaining exactly five seconds a day – this would be easy enough to calculate and compensate for, but it was rarely that straightforward. In 1840, Henry Raper, a British navy lieutenant and authority on navigation who normally extolled the virtues of chronometers, observed how 'Chronometers are generally found to perform best at the beginning of their voyage; many subsequently become useless from irregularity, and some fail altogether. They are liable, also, to change their rates suddenly, and then to resume the former rate in a few days.'

The rigours of the open ocean posed further challenges. Most early-modern navigational instruments, chronometers included, suffered varying degrees of inconsistency caused by the dramatic changes in temperature, salty air, humidity and even magnetism from the many iron goods on board a ship.*

To keep them protected, chronometers were kept in wooden boxes, which were susceptible to warping. This meant that when the time came to wind the chronometer it was sometimes comically impossible to open the box. Their winding and monitoring was restricted to experienced ranking officers, and the boxes were locked to prevent curious hands from tinkering with anything. The lock introduced the potential for that most inbuilt and fundamental of human errors – losing keys. Astronomer William Bayly, who served Captain Cook onboard the *Adventure* on his second voyage through the Antarctic and Pacific, reported several occasions when he had to rescue the ship's chronometer from incarceration. The first incident occurred when an officer accidentally bent one of the catches in the lock, so the catch needed to be sawn through and repaired. Shortly after, a key snapped off inside the lock. Then, a month later, Bayly had to break in a third time after someone left the ship with the key.

And no amount of temperature regulation or spare keys could guard against the grave threat of the ship's cat. One of the most famous cats in exploration history, Trim belonged to Captain Matthew Flinders (although, from my experience of living with cats, it is perhaps more accurate to say Captain Flinders belonged to Trim) and accompanied him on the first circumnavigation of the continent now known as Australia. Flinders writes how:

> Trim took a fancy to nautical astronomy. When an officer took lunar or other observations, he would place himself by the timekeeper, and consider the motion of the hands, and apparently the uses of the instrument, with much earnest attention; he would try to touch the second hand, listen to the ticking, and walk all around

---

* Although, I should say, Harrison worked hard to accommodate temperature change and among his considerable contributions to the history of horology are huge advances in our understanding of, and compensation for, temperature variation.

the piece to assure himself whether or no it might not be a living animal.*

Cats aside, the chief challenge to the viability of early chronometers was the cost.† Although prices reduced over the years, at the end of the eighteenth century chronometers still cost between £63 and £105, which when compared to the maximum annual wage of a Royal Navy lieutenant – £48 – was still completely out of most sailors' reach. And even when they could afford them, navigators continued to use the old methods to calculate their whereabouts – a chronometer alone was just not reliable enough. They were often used in conjunction with sextants, hand-held instruments that use mirrors to measure the angular distance between two objects. Sextants helped sailors to make celestial observations, which remained an essential means of measurement, especially when sailing into the unknown. They would also use them to check the chronometers' timekeeping against astronomical references whenever a crew could make land (and find a level stationary surface from which to make very precise measurements). This became an indispensable back-up for the chronometer, almost like a factory reset.

For me, Harrison's greatest achievement wasn't the creation of H4, or whether he won the Longitude Prize. (The Board of

---

* Trim, it appears, was a little more restrained than the ship's cat aboard HMS *Discovery*, which under the captaincy of George Vancouver set about mapping the west coast of North America between 1791 and 1795. Vancouver's exploration was assisted by a number of chronometers, an astronomical regulator clock, sextants and a pocket watch with a seconds hand. It was this pocket watch that unfortunately fell foul of a curious kitten at the beginning of the voyage. Excusing the mischievous cat for breaking one of the ship's state-of-the-art timekeepers, the exploration's astronomer William Gooch wrote that 'she is a very young cat & perhaps its beating attracted her notice'.

† These early chronometers were not cheap. In 1769, watchmaker Larcum Kendall was paid £450 to make the first replica of H4. Called K1, it took him two years to complete, and he received a further bonus of £50 when it was finished. That total of £500 was just under a fifth of the purchase price of the entire HMS *Endeavour*, which had been bought by the navy for £2,800.

Longitude, equivocating to the end about whether he met the terms of their prize, eventually gave him a further final award of £8,750.) It was demonstrating, for the first time, that a mechanical time-keeper – a watch, no less – had the capacity to solve one of the greatest problems of the day.

Despite its limitations, the chronometer was a valuable tool that, in combination with other methods, made charting the world possible for all. Not only did the chronometer help sailors to find their way through the vast open oceans, but those navigations enabled geographers to map the world more accurately. Imagine how dangerous it would have been to travel from, say, Portsmouth to New York without knowing exactly what shape the United States' coastline is, or where precisely the port you're heading to is located. The world as we know it today was made through the adventures and exploration of the eighteenth and nineteenth century – exploration that, in many ways, was made possible through advances in timekeeping.

But this expansion of our horizons had a dark side. Antiquarian horologists have not always acknowledged the role some of these inventions played in more damaging developments. When you read about the history of the Board of Longitude, the focus is usually on the preservation of the lives of sailors making perilous long-distance voyages, and on the benefits the chronometer brought to cartography. Little mention is made of the other interest Western nations had in perfecting transatlantic trade in the eighteenth century, and how improvements in navigation aided and refined the systematic enslavement of millions of Africans and their subsequent transportation to the Americas. There is rarely an acknowledgement of the impact being 'discovered' had on indigenous Australians, or how naval advantages assisted in the colonisation of India or South America. And although the Board of Longitude existed to support the Royal Navy, because chronometers were so expensive they were far more likely to be used by highly profitable commercial trading bodies like the East India Company, which relied on slave labour and trafficked slaves from East and West Africa. In 1802, thirty years after the invention of the chronometer, just 7 per cent of Royal Navy vessels had actually been supplied with chronometers.

The leaps in accuracy achieved for the chronometer made a different kind of watch – and time – possible. And for this, Harrison should share his crown with Mudge. While Harrison's legend overshadowed pretty much all of his golden age contemporaries, in fact no maker has left as lasting a mark on the advancement of the watch movement as Thomas Mudge. Mudge's lever escapement watches turned out to be more accurate and more reliable even than Harrison's, and he incorporated them into a chronometer just as Harrison won the Longitude Prize. Had Mudge succeeded a few years earlier, he likely would have beaten Harrison and claimed the prize, and lasting celebrity, for himself. Nevertheless, his legacy lives on in the watch. It is the descendant of Mudge's lever escapement that is still ticking on your wrist.

5

Forging Time

After all we are a world of imitations; all the Arts that is to say imitate as far as they can the one great truth that all can see. Such is the eternal instinct in the human beast, to try & reproduce something of that majesty.

Virginia Woolf, in a letter to a friend, 1899

In 2008, I was a cataloguer at an auction house in Birmingham. After graduating from horological college I had worked my way up from the bottom to become their chief watch specialist. The job was always varied and surprising. Some days I'd be dealing with infinitely precious privately owned pieces making a rare outing from their home in a bank vault, other days I'd be rifling through boxes of jumbled junk, trying to identify anything with potential value. One morning, in a box of antique silver tableware, I came across a silver pair-cased watch with a hallmark dating it to 1783. The inner case housed the mechanism and had a hole in the back for winding. On the front, the dial was visible under a domed bullseye crystal, named after its large eyeball-like profile. This all sat snugly inside an outer, more durable case that protected the delicate mechanism from the elements. I held the watch up to my Anglepoise light and looked through my eyeglass. The dial was an ornately engraved sheet of silver, with inlaid black wax numerals – a technique known as *champlevé*. Its centre had been pierced with a delicate acanthus-leaf scroll detail to reveal the bright flash of blued steel shim underneath. Just above the centre of the dial, framed in a decorative scroll, was the watchmaker's name: John Wilter.

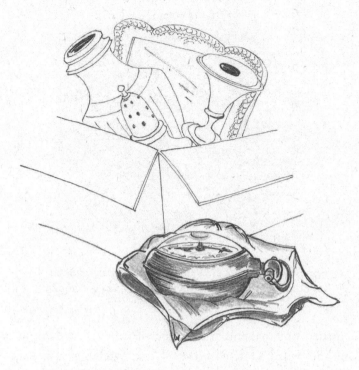

A cardboard box of silverware brought to the auction
house for valuation and cataloguing.

So far, I hadn't spotted anything hugely noteworthy. I removed the
outer case and opened the inner case to reveal the movement, which
was gilded brass and also engraved and decorated with areas of pierced
detail. It too was signed 'John Wilter, London'. My curiosity was now
piqued. The design was highly unusual for an eighteenth-century
English watch. The dial was arcaded (meaning that the minute track,
which is normally a perfect circle, had scalloped arches surrounding
each of the numerals) – a motif popular in Dutch clocks of the time,
but almost unheard of in England. The movement's components were
likewise crafted in a continental style and were of a lower quality than
you would expect to see in genuine London work. I pulled my bible
– Loomes' *Watchmakers and Clockmakers of the World* – off the shelf and
leafed through its pages until, eventually, I found the relevant entry:
'Wilter, John – perhaps a fictitious name'.

An arcaded *champlevé* watch dial signed 'Wilter, London'.
The acanthus scroll work in the middle was skeletonised
to reveal blued steel underneath.

It turned out that this was a type of watch I'd never heard of before: a so-called 'Dutch forgery'. These watches were low-quality fakes that most horologists either ignored or condemned. They typically purported to be English watches but were Dutch in style. Why, I wondered, would someone be forging an English watch in the Dutch style? And who was John Wilter? I found no evidence that anyone, let alone a watchmaker, with that name even existed at the time. I didn't know it then, but I would spend a whole decade of my life searching for John Wilter.[*] My quest would teach me a great deal about how watches became accessible to all.

I began my research at the Horological Study Room of the British Museum. I've spent countless hours there volunteering and researching; in many respects it's my spiritual home. To get there I had to weave my way through throngs of tourists browsing the galleries,

[*] John Wilter and Dutch forgery watches became the subject of my PhD.

and a muffled hum of voices in other languages. I confess, I always loved being watched by these curious visitors as I sidestepped the security barrier and unlocked the imposing oak double doors that led out of the gallery. It's strangely empowering to open a door that's many times taller than you.

The marble floor of the British Museum's Great Court is inscribed, in jet-black stone, with a quotation from Tennyson: 'And let thy feet millenniums hence be set in midst of knowledge.' For a tactile, object-obsessed thinker like me, the British Museum really is a temple of knowledge. No matter how many times I opened those doors it felt like I was opening the gates to a secret world – a vast network of hidden treasure.

Great museums are like icebergs, with only a tiny fraction of their vast collection visible to the public. As I closed the grand doors behind me with a reassuring thud, I left the tip of the iceberg behind me. The acoustics immediately changed. I walked down long, quiet corridors, flanked by floor-to-ceiling cases filled with antique books, behind rippled antique glass. As I descended into the basement, the temperature dropped. The walls were now covered with bright white tiles, an aesthetic somewhere between a London Underground station and a Victorian hospital. Here, between cabinets of Bronze Age pottery, was my destination: an inconspicuous door with a buzzer. I was met by the curator and escorted into a room lined with grandfather (or longcase, to use the correct name) clocks where, at the far end, were two long rows of mahogany cabinets positioned back to back. Those cabinets housed the museum's 4,500 watches, arranged in hundreds of specimen drawers. The collection spans the whole history of the watch, from its invention in the sixteenth century to the present day. It contains the work of almost every well-known maker as well as pieces made by artisans whose names have been lost to history. It also contains dozens of Dutch forgeries.

The horology collection is the only collection in the museum where conservation work is carried out by its curators, as horology is one of those rare subjects where theorists have to be practitioners in order to care for the objects of their study. I've found several kindred spirits among these curator-conservators. When I started volunteering in 2008 the then head curator David Thompson, who would go on to supervise my PhD, became my informal mentor. David had studied at the old Hackney College Horology School, which closed in the late 1990s, and spent thirty-three years working with the British Museum's collection. I owe David my taste for the horological hunt. His office was a desk hidden at the end of a warren of crowded bookcases filled with centuries of horological literature. He had a librarian's memory. If confronted with a question to which he didn't have an immediate answer (they were rare), he could instantly locate the relevant information from the shelves around him. I now model my own study on his office.

When he retired, Paul Buck took over. Paul is, to me, one of the most fascinating people on the planet. Every single time I speak to him, he tells me something that makes me gawp in amazement. That's no mean feat in an industry populated by what my husband calls 'woolly-hatted, cream-horn-eating, walking-sandal-and-sock-wearing garden-shed-tinkerers'. We watchmakers are not generally 'cool' people. We're the engineers of a bygone era who spend most of our day indoors working on eye-strainingly small objects with little to no human contact. But Paul (AKA Pablo Labritain, drummer in the punk rock band 999) is an exception. His specialist area is old cuckoo clocks (properly termed Black Forest clocks), but for many years he spent his lunch break practising the drums in the museum's Radium Room, where any objects containing radioactive substances are safely contained in steel chemical storage cabinets with extraction. I'm not sure anything could be more 'punk' than playing drums in a radioactive room.

Aside from his skills as a drummer, Paul is an exceptional restorer and teacher. It was Paul who taught me the painstaking process of repairing miniscule fusee chains. He showed me how to fashion the end of a round needle file into a three-sided cutter to cut through the rivet holding the tiny pins that secure each link in place. He taught me how to remove the pins and the damaged links either side of the break before threading the severed ends back together. He told me to use the steel of a sewing needle, as opposed to modern carbon steel, as his many years at the bench had taught him this better replicated the metal original watchmakers would have used centuries ago.*

Paul and David gave me permission to dismantle the Dutch forgeries. Like a forensic scientist, I used a combination of microscopic, X-ray and X-ray fluorescence scanning to build up a picture of their lives. All of them, like Wilter's, were verge watches.

When I first restored a verge watch as a student, I asked my tutor what sort of timekeeping I should aim for. 'Getting them running at all is a triumph,' he replied unencouragingly. Many modern repairers refuse to work on them. For one thing, they have frequently suffered from centuries of bodged repairs, which need to be carefully undone before the watch can be properly restored. I once handled a Dutch forgery whose outer case was adorned with a heavily worn depiction of the abduction of Helen of Troy in *repoussé*.† A previous restorer, however, had haphazardly re-engraved Helen's missing features, making her look less like the face that launched a thousand ships and more like Munch's *Scream*. The movements, meanwhile, are often clogged up with rust, dust, or bits of cardigan, all of which need to be painstakingly removed. The bearings are typically made of brass, as opposed to the harder-wearing materials like ruby used in higher-quality and more modern watches, so the mechanism wears itself out as it runs. As the movements weren't standardised, no spare parts are

---

* When Paul retired, although he's still touring with 999, the department was handed down to Oliver Cooke and Laura Turner, to whose continuing patience and support I am indebted.

† A relief design on a thin sheet of metal hammered through from the underside.

available; you can't even pinch a component from another watch of the same age without laborious customisation. Anything broken needs to be remade by hand. But I can't help loving these watches. Every single one of them is a character. Each one has its own clumsy idiosyncrasies, like an old car or a favourite pair of jeans that are falling apart but you can't bear to part with.

It was this handmade element, however, that initially made watches so expensive. Watches were complicated and time-consuming to produce: thirty or more individuals, each with different, interconnecting skills, would have been involved in the process of making a single watch. As a result, even the largest eighteenth-century British workshops only had the capacity to make a few thousand watches a year. And yet as the century progressed, a new, more affordable kind of watch began to appear in pawnbrokers' windows and on market stalls. At first, they appeared only in ones or twos, but by the end of the eighteenth century they far outstripped the numbers that Britain's established watch industry was capable of producing. Someone, somewhere, was making them more quickly and cheaply.

London's watchmaking world was, as we have seen, something of an elite boys' club. The long and expensive apprenticeships required by the Clockmakers' Company meant that training in the capital was only an option for the privileged few, which in turn placed a stranglehold on the City's watch production. This encouraged them to turn to the trade further afield to assist with their production. To keep up with the growing demand for watches, their makers became increasingly reliant on buying in what we now call *ébauches* (blank movements supplied ready to be finished and branded by another watchmaker), made in regional workshops in Lancashire and later Coventry, by artisans who had never served a formal apprenticeship. One worker in the north said it was 'only those who intended to become masters' who served their time formally while 'the rank and file of the workpeople never became formally indentured'.

Yet accounts from Prescot in Lancashire, where some of the finest horological tools together with parts and full *ébauches* were made, describe an exquisite level of craftsmanship achieved by non-apprenticed makers. One observer noted how these craftsmen formed miniscule 'bay leaf'-shaped gear teeth by eye despite the fact that they 'would stare at you for a simpleton to hear you talk about the epicycloidal curve'. This form of watchmaking was often a side hustle. Farmers who owned enough land to feed themselves and their families but had little left to trade for profit often engaged in other enterprises like spinning, weaving and, indeed, watchmaking to augment their income. Some local manufacturers set up workshops on small farms, running them in conjunction with their main business. These often beautifully executed watch parts would then be packed up and sent to watchmakers in London and across Britain, where they were finished and made up into watches.

British watchmaking capacity and ability to compete on prices was also limited by the exclusion of women from the workforce. Artisan culture was almost exclusively male. Some women were listed as masters and apprentices in the Clockmakers' Company, but most of them were actually milliners (milliners didn't have their own company at the time so were listed under others). The number of women serving formal apprenticeships was incredibly low across all trades, at just 1–2 per cent. One ongoing study has so far found only 1,396 women associated with watchmaking in the United Kingdom between the seventeenth and twentieth centuries: that might sound a lot, but when you consider that in 1817 alone there were over 20,000 people involved in watchmaking just in London, you realise it isn't.

Elsewhere in the world, for example in Switzerland, and later in the US, women were being welcomed into the workshop with open arms. As much as I would like to say this was an act of workplace equality, in reality women could be paid less, meaning the watches they made could also be sold for less.

There still aren't many female watchmakers. I was – and am – a rarity in the field. This has been a mixed blessing for someone who suffers from anxiety and a chronic sense of imposter syndrome. Being different makes it easier to be noticed. And being noticed comes with benefits and drawbacks. I have been lucky to have support from amazing friends and mentors, without whom it's unlikely I would have even finished my training. But I've also taken a lot of flak. In my very first workshop there were some who made it clear they thought I'd only got on the course because of tokenism. I once had a tutor petition an employer who'd offered me a summer placement to withdraw it and offer it to one of his male students. I've heard others say there's no point training women to be watchmakers as we have children and give up the profession. I've been told on several occasions, 'You're not special, you know.' This always makes me reflect: how could anyone think that I believe myself to be special? I suspect I'll never stop feeling like an outsider.

One morning in the car on my way into the workshop, I was listening to an interview on *Woman's Hour*. Cambridge scholar Morgan Seag was discussing the reasons why the British Antarctic Survey banned women from visiting the Antarctic until 1983. There were lots of predictable excuses: it was another era; they felt women wouldn't be interested due to the lack of toilets, shops and hairdressers; they were concerned about the impact of releasing females into a male colony. But what really struck me was her description of ice as a kind of stage of masculinity, which women, men feared, would undermine. Pioneering explorers like Robert Scott and Ernest Shackleton had created an illusion of the hero, of brave men tackling terrifying crevasses, extreme weather and starvation, to boldly strike out into the unknown. Professor Liz Morris, the first woman to work in the interior of the continent in 1987–88, said there were men who were resistant to her participation and quoted US Antarctic Leader George J. Dufek's view that if 'a middle-aged woman with no particular physical skills could hack it, then how could they [men] be heroes?'

It suddenly dawned on me that the same thing was happening in my field. Young watchmakers are also raised on the stories of heroes

– Huygens, Tompion, Graham, Mudge, Harrison – the geniuses of the golden age who designed extraordinary works of engineering to solve some of the greatest scientific problems of their day. Men – and they *were* all men – who socialised with aristocracy and wowed audiences with their mechanical ingenuity, created objects out of tiny pieces of metal that moved independently, as if by sorcery. The mechanism of a watch has been perceived by society as something so perfect and complex that it was used in an argument to counter the existence of God (usually assumed to be another man). When the Clockmakers' Company was founded in 1631, its charter declared that it would oversee the 'Fellowship of the Art and Mysterie of Clockmaking'. To this day, watchmaking is still viewed as a dark art that only a special few can understand. And yet, here I am: a socially awkward, tattooed woman, raised in a working-class household, with absolutely nothing 'special' about me – and I'm a watchmaker. If someone like me can become a master watchmaker, anyone can.

No wonder I found the dodgy John Wilter so appealing: I was an underdog drawn to another.

In the British Museum, I found the same *ébauche* makers' marks (called platemakers' marks and hidden under the dial) on movements signed with a range of fictitious names. It seemed that a relatively small number of workshops were making huge numbers of watches on a scale unseen in England at the time. Not even the most active workshops in the north of the country could match this pace. At first sight, the watches looked Dutch: the scalloping of the minute track and the shape of the balance bridge, the component that secured the upper pivot of the balance wheel, were stylistically Dutch rather than English in design. But the quantities didn't make sense. Although there were a number of very talented watchmakers in the Dutch Republic at this time, the Dutch industry was tiny in comparison to London's – and nowhere near big enough to produce the vast the number of 'Dutch forgeries' entering the market.

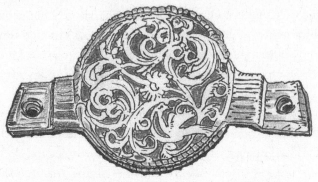

A two-footed balance bridge. This design is different to the
single foot normally found in eighteenth-century English
watches but was popular among watchmakers on the Continent,
including the Dutch Republic and Switzerland.

By studying the marks hidden in these watches, and comparing
them to archival records, I was able to work out what was happening.
Dutch merchants were indeed commissioning them in their national
style but, knowing that consumers preferred London-made watches
in the same way we might prefer German cars, Japanese cameras or
Belgian chocolate today, they signed them with English names, in the
hope of fetching higher prices. But, curiously, these fakes weren't
being made in the Dutch Republic or England. The assay marks,
hidden signatures, contemporary witness statements, newspaper
reports and even accounts from the places where they originated
made it absolutely clear: these watches were coming from Switzerland
where, since the start of the eighteenth century, a new approach to
watch production was being perfected. It was known as *établissage*.

If traditional watchmaking relied on small collectives of skilled
artisans passing items between different workshops, and Britain's
unofficial watch trade relied on cottage industry workers, *établissage*
brought a larger number of workers under a single roof, known as a
manufactory. In the manufactory, labour was organised as a production line with specific chainmakers, springmakers, wheel-cutters and
pinionmakers working alongside each other. Although the techniques
and equipment were much the same as those used in traditional
manufacturing methods, the *établissage* system dramatically

streamlined production under the management of a single firm. It was highly efficient and, as a result, manufactories could create huge numbers of watches. Where Britain's largest workshop could produce a few thousand watches a year, a Swiss manufactory could produce 40,000. This completely revolutionised the industry. As a result of *établissage*, European watch production rose dramatically over the course of the eighteenth century, reaching an estimated 400,000 per year in the last quarter of the century, possibly even more.

Switzerland's geography made it perfectly placed to supply the trade. The country was located on a major trans-European trade route, with Dutch, French and English merchants constantly passing through as they made their way from the River Rhine to the River Rhône, which acted as a natural transport link between the Baltic Sea in the north and the Mediterranean in the south.* The frequency with which watch manufactories appear in the linking land route between the two rivers is a strong indicator of a merchant-directed industry. Merchants, who were constantly travelling across Europe as

---

* I've read a number of stories about how these watches made their way out of Switzerland and around the world. The market for these Swiss-produced fakes was shadowed by a criminal underbelly willing to smuggle them to their final destination. Watches are small, making them easy to transport in large quantities, tucked away in trunks under linen, in empty wine barrels – or strapped to hungry dogs. In 1842 the director of French customs reported that his officers at the border were being set upon by packs of ferocious dogs, in a state of 'madness'. The dogs had apparently been taken across the mountainous Swiss–French border, deprived of food and beaten before being sent out into the night with watches strapped to their bodies. Laden with up to 12 kilograms of watches each, the dogs would dash back across the border straight to their master's house, where food and good treatment would await. I have to say I'm somewhat sceptical that 'wild' dogs could be trusted to traverse miles of mountainous terrain between abusive masters like masochistic carrier pigeons. My own friendly and moderately trained Staffordshire Bull Terrier cross, Archie, cannot be trusted to make it from one end of our workshop to the other without finding a distraction. The bulk of the hundreds of thousands of watches leaving Switzerland each year would have been shifted via more reliable methods – in carts on little-trodden mountain passes and with the ships and merchants trading across Switzerland between the Rhône and the Rhine.

well as further afield, were much more aware of changing fashions, and much more in tune with market demands, than overworked craftsmen stuck in their proverbial cottages. This led to a paradigm shift in watchmaking: one in which merchants told craftsmen what to do, rather than simply retailing on their behalf.

As I rifled through the British Museum's collection, I was constantly on the lookout for evidence of the Dutch forgeries' fictitious origins. I found several more John Wilter watches. Curiously, a small number of these were very high quality and English in style, but the majority were lower-quality Dutch-style forgeries. I also found all kinds of spelling mistakes on the movements, like the suspicious typos you often see in junk emails. Renowned father–son watchmakers Joseph and Thomas Windmills had been credited as 'Wintmills, London' on one watch and 'Jos Windemiels, London' on another. I also found 'Vindmill', 'Wintmill', 'Windemill' and 'Vindemill'. A Jonh Wilter and a John Vilter also featured, although, unlike the Windmills, the identity of the John Wilter they were apparently imitating remained a mystery to me. Nevertheless, as Dutch-speaking merchants commissioned English-named watches from French-speaking manufactories in Switzerland, something was clearly getting lost in translation.

But Dutch forgeries were immensely important. These watches, which often undercut their genuine London competitors by more than 50 per cent, represent the first mass-produced timekeepers. This was the moment when portable time ceased to be the preserve of the ultra-rich. These watches made no contribution towards the accuracy or reliability of the watch, nor were they technically or aesthetically innovative, but they were cheap – and that's what makes them interesting. For the first time since their invention, a way had been found to make watches affordable. By the end of the eighteenth century they were becoming an increasingly common accessory among ever wider social groups.

My Wilter watch is evidence of one of the most significant socioeconomic developments of the eighteenth century: imitation. From the

1760s onwards the Industrial Revolution gave rise to an emerging middle class with aspirations beyond their financial means. Theatres, parks and the emergence of free museums and art galleries created more opportunities for the wealthy and the aspiring to collide. And as literacy rates improved and print production increased, newspapers provided these people with a window into the lives and possessions of the upper classes. It drove the desire for luxury possessions. And as these material luxuries remained out of financial reach for most of the population, the solution was to fake them.

From painted blue Oriental-inspired ceramics to plated metal and steel cut and polished to look like diamonds, an entire industry rose up, producing pseudo-luxuries for a new and rapidly expanding group of people. Sheffield Plate,* invented around 1742, became a highly popular commodity for any committed social climber. Under the dim candle-light of a Georgian dinner party, and after a few too many glasses of wine, guests could easily be tricked into thinking their supper was being served on an exceptionally expensive solid silver service. Similarly, ormolu – the name given to gilded bronze or brass objects – might convince visitors that their host's home was filled with solid gold *objets*. Georgian industrialists made keeping up with the Joneses possible on almost every level – provided you didn't look too closely.

Dutch forgery watches were just one part of this larger process. In the eighteenth century, watches – which were worn prominently on chatelaines, an ornate style of chain that hung from a waistband – were conspicuous signifiers of wealth and status. So much so, in fact, that in 1797 the prime minister and chancellor of the excheq-uer William Pitt introduced a tax on watch ownership. He justified it by declaring that owning a watch was a mark of luxury and so proof that the owner could afford to bear the additional tax.

---

* Formed by rolling and pressing a thin sheet of silver onto a much cheaper copper base metal, plate could be used to create everything from candlesticks and other tableware to complete dinner services. To add to the web of illusion, makers' logos and marks were commonly designed with striking similarity to genuine hallmarks.

(Needless to say, the tax was exceedingly unpopular. The middle class rebelled, with some going as far as scrapping the gold cases on their watches and having new ones made in cheaper metals to avoid their watches being classed as eligible for taxation.)

If a century earlier most people's lives were lived by a communal clock, personal watches were now everywhere. We see them in visual art. Clocks and watches make regular appearances in the work of William Hogarth, who used them to track his characters' progress through his stories. His famous series *A Harlot's Progress* charts the corruption of an innocent country girl, Moll Hackabout. The closer Moll gets to her eventual demise (incarceration and then death from venereal disease), the closer the clocks in his engravings edge towards the eleventh hour. Interestingly Moll, a 'harlot', apparently owns a fancy 'repeating' pocket watch.

The proliferation of watches was matched by a surge in pickpockets and street robbers. Watches are the pickpocket's loot of choice in John Gay's *The Beggar's Opera* (1728). Timepieces were highly prized by thieves, commonly traded for services like prostitution or to pay off gambling debts, as happens in Plate VI of Hogarth's *A Rake's Progress* (published 1735). Records from the Old Bailey show that inns, taverns and gin booths were the preferred hunting grounds for pickpockets, with thefts peaking between 8 and 11 p.m. (when nearly half of all reports were made), dropping off after midnight, then rising again in the morning, around 7 a.m., when people woke up with a sore head to discover their watch was gone. Stolen watches were quickly and easily disseminated into London's criminal underground through pawn-brokers and second-hand shops. Unscrupulous jewellers readily took them in; there were even horologists willing to modify them by changing names or hallmarks to avoid them being identified by their true owners. Daniel Defoe's irrepressible Moll Flanders in his 1722 novel, one of the most famous pickpockets in literary history, regularly uses a watch we can assume was pinched from an unsuspecting victim.

The Old Bailey's papers don't simply reveal increased levels of watch ownership and theft, but also an increased level of time awareness. Over the course of the eighteenth century, the specific time of an event became a more and more common part of witnesses' testimonies in crime reports. Thomas Hillier, who appeared in court in 1775 after his silver pair-cased watch was stolen by highwaymen between Hampstead and London, stated that the event occurred at 'about a quarter after nine at night' and the whole ordeal lasted around 'a minute or a minute and a half'. Although Hillier's account was unusually specific, it was one of thousands from Londoners in this period that had begun to give times, dates and durations of occurrences. Time awareness was slowly but surely growing.

The story of eighteenth-century watch-time has two very different sides. On the one hand, you have the glamorous golden age of horological advances, of marine chronometers, of watches crafted by some of the most highly educated scientists in society. But you also have a murky underbelly of fakes and forgeries that, in my view, is no less interesting and important. Dutch forgeries disrupted the connection between artisan watchmakers and wealthy patrons. They made a giant leap towards making watches affordable, paving the way for other later companies to make them truly accessible to all. For that reason, to me they are as significant in the story of horology as Harrison's chronometers. In the end, how can we argue an innovation is truly world-changing if only a small elite can access it? By widening access to time, cheap watches helped to close the divide between rich and poor, the aristocracy and the masses. They democratised time.

But what about John Wilter? Was he too an invention, a fake? Several years ago, I happened to be leafing through the minutes of an 1817 hearing at the House of Commons when I finally found a contemporary reference to him, from a person who claimed to have known the man behind the myth. I remember when I saw that name on the page – a name that had haunted me for years – I froze for a moment. I looked away, took a deep breath, and started reading. The witness – another watchmaker called Henry Clarke – spoke admiringly of a man who he claimed had, by then, passed away. He'd been making watches to commission for a merchant who had ordered that:

> [He] introduced the making of watches with the feigned name of 'Wilters, London,' on them; those watches were well made, and would have done credit to the maker, who should have put his name upon them; other persons speedily imitated the external appearance of the watches . . . [but] those had sham day of the month, dials and hands without and wheels to move them, and also the sham appearance of being jewelled in the pivot holes . . . The last I saw of those spurious watches were offered to me for sale at 34s. each, but really were good for nothing; whereas the first introducer of watches, with that feigned name, was not overpaid at eight guineas each.

Wilter, I discovered, was both real *and* fake. The name itself was probably the invention of a Dutch merchant who wanted something that sounded English but also couldn't be traced. But the man he originally commissioned *was* a genuine English watchmaker of considerable skill. 'John Wilter' became something of a brand. The merchant had then realised he could boost his profits by having Wilter watches made cheaply on the Continent. This tied up perfectly with the evidence in the British Museum, where I'd found a few high-quality Wilter watches as well as several more typical Dutch forgeries. This short passage filled in a gap I hadn't previously been able to explain. And these watches are a prime example of why fakes shouldn't be scorned, as the value they can add to our understanding of the world, of industries and even economies, is immense. As a scholar I studied them, and now I've even managed to collect a few of my own, including my most cherished ones, by my infamous John Wilter.

6

Revolution Time

At times the heart plays tricks and lets us down. The vigilant are right. For God – the mighty Breguet – gave us faith, and seeing it was good, improved it with a watchful eye.

Victor Hugo, *Les Chansons*, 1865

In August 2006, Rachel Hasson, the artistic director of the LA Mayer Museum of Islamic Art in Jerusalem, received a call from Zion Yakubov, a watchmaker in Tel Aviv: a cache of stolen antique timepieces had been found and he thought she should come and take a look. She'd had calls like this before; they were always a hoax. But this one was different.

Twenty-five years earlier, the museum's unparalleled watch collection had been stolen in a heist that had confounded the police, the Israeli Intelligence Community and even Samuel Nahmias, Israel's most accomplished detective. On the night of 15 April 1983, 106 irreplaceable watches and clocks, with one particular piece valued at more than $30 million, vanished. A huge search began but every line of enquiry yielded nothing. As the years passed, it seemed as though the objects had disappeared from existence.

Detectives had followed leads as far afield as Moscow and Switzerland, but it turned out the watches were in a storage facility in Tel Aviv, just an hour away from where they were taken. Hasson and Eli Kahan, a member of the museum's board, travelled to the office of Hila Efron-Gabai, a lawyer who had been asked to return the watches anonymously for her client. They identified the watches from their serial numbers; some were intact, others were damaged. But one watch brought tears to Hasson's eyes. There, wrapped in yellowing newspaper, lay 'the Mona Lisa of watches', made by Abraham-Louis Breguet for Marie Antoinette, the most complicated, beautiful and valuable watch ever made.

Police finally tracked the robbery to notorious Israeli thief Na'aman Diller, who had confessed the crime to his wife on his deathbed. They were disappointed not to have a chance to interview Diller, who had 'a unique style'. (In 1967 – with a short break to fight in the Six-Day War – he dug a 300-foot trench at the back of a Tel Aviv bank in order to blow open a vault.) Diller always acted alone. This time he had used a hydraulic jack to force the museum railings apart and a rope ladder and hooks to climb 10 feet to a small window just 18 inches high. Diller, who was 'whippet-thin', then slithered through the narrow opening before making off with most of the priceless horology collection.

This is one of the more recent chapters in the story of the 'Queen's watch', which began some 200 years ago.* In 1783 – the same year my humble John Wilter forgery was assembled – an anonymous admirer of Marie Antoinette sought out the services of the most famous watchmaker in Europe in order to commission a very special gift. No doubt the admirer hoped to gain favour with the queen, who was already an enthusiastic patron of Breguet's work. The commission was for a watch more complex than any that had gone before it. It was to include all the most advanced and complicated mechanisms of the day and no expense should be spared: gold should replace other metals wherever possible, including in the mechanism itself. It was to be a watch for its times, exemplifying the best and worst of the *ancien régime*.

The best: the lavish tastes of Louis XVI's court meant there were no boundaries on creativity. To make this watch nonpareil, Breguet

* Marie Antoinette's watch, like Queen Charlotte's Thomas Mudge watch, is commonly referred to as the 'Queen's watch.'

was given an unlimited budget, and as long as he needed. It is hard to express how exciting it would be, for a maker, to be presented with an open chequebook *and* an open calendar – pure scope in time and budget to test your skills and ingenuity to their limits. Even for the successful watchmaker – and by that time Breguet had been appointed *Horloger du Roy* to Louis XVI – cashflow is one of the hardest things to manage. An exquisite, handcrafted timepiece costs an obscene amount of money to buy and to make. For us today we can sometimes expect a six-figure sum – but by the time you've deducted overheads, taxes, material costs, outworkers' bills and divided it by the x number of years it's taken to make, it can sometimes barely cover a living wage for its maker.

For Breguet, the unlimited time allowance and budget enabled him to pour everything he'd spent his lifelong career mastering into a single piece. The watch had a total of twenty-three complications, those functions which are surplus to telling the time. It was self-winding and could strike the time out loud, sounding the hours, quarters and minutes on finely tuned gongs made from wire, and it displayed the equation of time. It had power reserve indication (it could run for forty-eight hours from full wind), a chronograph, a thermometer and a perpetual calendar à la Mudge. In total, the watch required 823 parts squeezed into a 6-centimetre diameter pocket watch and is still considered one of the five most complicated watches in the world. Its engineering was so spectacular (and exquisite) that the whole movement was left visible through the glass dial and casing to show the mechanism busily working away inside.

The worst: while Breguet worked on this infinitely luxurious, money-no-object watch, people were starving. France, still locked into a feudal system, taxed the peasants (96 per cent of the population, who lacked any political or economic power), while the clergy and aristocracy reaped the benefits. From 1787 to 1789 France weathered terrible harvests, droughts, cattle disease and skyrocketing bread prices. Meanwhile, the French government was bankrupt – their costly involvement in the American War of Independence, and Louis XVI's extravagant court, had taken a heavy toll on the national purse.

Tax increases lit the touch paper, the outrage of the citizens of Paris was unleashed, culminating in the French Revolution in 1789.

Marie Antoinette never got to see her watch. Breguet's workshop was disrupted by world events and it was not completed until 1827, thirty-four years after its beautiful, if politically insensitive, intended recipient had met her fate at the guillotine.

True master watchmakers were, and still are, rare creatures. In his entry on horology in Diderot's *Encyclopédie*, the celebrated watchmaker Ferdinand Berthoud (1727–1807) describes the demands of his craft: 'thorough mastery of *horology* requires *the theory of science, the skill of handwork, and the talent for design*, three qualities that are not easily fostered in the same individual'. I like his admission that it is hard to unite these qualities in the one human. Craig and I have always said that individually we're good watchmakers, but together we make one incredibly good watchmaker. We have complementary strengths: Craig is the illustrator, designer of the aesthetic, whereas I'm more mathematically minded and can turn his beautiful hand illustrations into precise technical drawings that we can work to. Craig has an incredible level of dexterity working with fine parts, like repairing damage to hairsprings, while I like finishing – putting the polished angles on the spokes of gear wheels, springs and plates by hand with tiny files. I love wheel cutting; positioning and cutting the hundreds of teeth in a watch by hand. I like the way you get into a repetitive, almost hypnotic motion with the cutter as you wind it forth and back, again and again, over and over. Craig finds it monotonous, but that's okay, because he has far more patience than I do for turning tiny balance staffs, just a few millimetres in length, by hand and polishing them to a hard, bright, mirror finish. In Breguet, however, all of those skills and more were uniquely combined.

In horology, it is something of a cliché to confess that your hero is Abraham-Louis Breguet. He is one of the greatest and most celebrated minds in the history of our industry. He started his apprenticeship in Paris in 1762 at the age of fifteen and, by his death in 1823, he had revolutionised his profession. The barrister and horology buff (and one-time owner of the 'Queen's watch' together with many others that were stolen from the LA Mayer Museum) Sir David Lionel Salomons famously declared that 'to carry a fine Breguet watch is to feel that you have the brains of a genius in your pocket.'

Breguet's extraordinary technical skill and inventiveness is the stuff of horological legend. No other watchmaker created as many inventions that have remained in use to the present day. He was the originator of the *perpétuelle*, the first automatic winding mechanism, which powers the watch through the movement of the wearer using the swinging action of a weight inside the watch to power up the mainspring. He invented gongs made from wire in his repeating watches that chimed the hours and minutes, allowing for these watches to be much slimmer than the traditional bells that existed beforehand. He developed what is still referred to as the Breguet overcoil hairspring, which was a technical improvement to the flat spiral design that preceded it. Breguet found that raising the last outer turn of the flat spring above the rest significantly improved isochronism, or the equal duration of each breath, and consequently timekeeping and accuracy. As before, all these developments are still featured in watches to this day; some have been advanced further, but others, like the overcoil, have remained virtually unchanged for over 200 years. Thanks to his innovations, watch movements became slimmer. Gone were the pebble-like pair-cased pocket watches of old. Breguet's watches were slim enough to fit in the tailored pockets that were becoming fashionable among gentlemen.

Ironically, for a man who lived in an age synonymous with excess, Breguet stripped his designs right back. His work was restrained and embodied purity of purpose. The pared-back yet elegant watch hands he used are still named 'Breguet hands'. Often in electric blue steel, or in gold, they are exquisitely long and slim with a small hollow circle near their tip, and a little arrow-like point to precisely indicate the hours and minutes on the dial. He favoured a style of engraving known as engine-turning. Instead of hand-engraving the precious metal of his watch dials, he used manually operated rose and straight-line 'engines' to create intricate geometric patterns. These machines are, in my opinion, among the most beautiful ever invented. Rose engines slowly rotate a series of different-shaped discs, translating this movement into a pattern that gets cut into the surface of the watch dial.* You could compare the process to a bicycle with, say, an octagonal wheel. As you ride along, the seat will bump up and down according to the shape of the wheel. In a rose engine, the bicycle seat is your watch dial, which moves around against a stationary cutter. By moving the dial past the cutter many times, and altering the cutter's position slightly each time, an orderly pattern is inscribed into the metal. They are remarkable, and hypnotic to watch in action, and can produce a huge range of patterns, from the basket weave, which looks a little like a tiny chequered chessboard, to rosettes, which radiate from the centre in different shapes, like ripples across the surface of a pool escaping a dropped stone. They can engrave milled circles, which are used as bands to border the numerals on the dial or embellish the main band of the case with a pattern a bit like the edge of an old pound coin. The effect, at once detailed and discreet, gives the dial a subtle lustre, the height of low-key luxury. The machines are now as rare as hen's teeth. I've occasionally commissioned this kind of work from specialists, but even if I *could* get my hands on my own machine, I would rarely use this style. It is so associated with Breguet, it feels like trespassing.

Breguet's watches also revealed their own magical mechanical workings. Instead of the classic full plates – two round discs sandwiching (and concealing) most of the movement – he dissected these plates into a skeleton-like series of 'bars', which allowed you to see into the movement and watch the train wheels (which gear down the power from the mainspring) turn and the escapement ticking

---

* Straight-line engines use a similar principle but move up and down rather than round in a circle.

through.* Breguet's movements were still made from brass gilded with yellow gold but, instead of the delicate floral and foliate engraving that would have typically embellished them, he chose to *frost* his movements, a satin finish that is both iridescent and yet matte, depending on how it catches the light. It can be achieved by various means, from the application of acid to the use of a stipple brush or, as watchmakers do today, silicon sand blasting (our silicon blasting machine was retired from a dentist, who was using it to clean the residue off tooth moulds.)† As counterfeiters started to mimic Breguet's designs, he developed a minuscule secret signature that was etched into his dials and was near impossible to replicate. You needed a magnifying glass to see it properly.

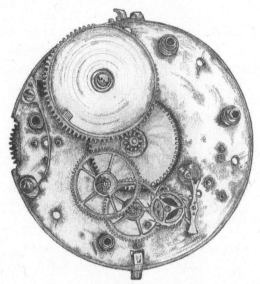

Train wheels running between the mainspring barrel at
the top, and the escapement on the lower right.

---

* It's worth noting that Breguet wasn't the inventor of this movement design, which had been popularised in the years preceding his rise to celebrity – and he never professed to be. But his innovations and notoriety have led to it becoming iconically 'his'.
† There is an overlap between dentistry and watchmaking tools. We both have a lot of pliers and we're used to working in small, fiddly places. In fact, between the seventeenth and nineteenth centuries some watchmakers used to offer dentistry – but not, for all his talents, Breguet.

Breguet's watches were precision scientific instruments, but many of these design quirks serve no functional purpose. They existed purely for delight, and joyously elevate engineering to become a work of mechanical art.

It was Breguet's stepfather, Joseph Tattet, who came from a family of watchmakers in Paris, who first introduced him to the profession. Breguet, born into a Huguenot family in Neuchâtel, Switzerland, in 1747, was the eldest of five and the only son. He lost his father at just eleven years old, and left school two years later. His mother remarried, and in 1762 young Breguet, like so many artisans before him, made his way across the 'permeable frontier' of the French–Swiss border to begin his training in his stepfather's family workshop in Paris.

Breguet attended evening classes in mathematics at the Collège Mazarin and his tutor, Abbé Joseph-François Marie, praised his talents in court circles so highly that word of his emerging talent made its way to the king and queen of France. Like George III across the Channel, King Louis XVI of France was fascinated by the mechanics of watchmaking. At the age of fifteen Breguet began an apprenticeship to a master watchmaker at the palace of Versailles.

When he was twenty-eight, Breguet founded his workshop on the ground floor of number 39 Quai de l'Horloge, the watch- and clockmaking quarter of Paris's bustling Île de la Cité. This was, to a great degree, made possible through the dowry he received from the wealthy and respected Parisian family of his wife-to-be, the 22-year-old Cécile Marie-Louise Lhuillier. Watchmakers often married later in life − in part because liaisons of any kind with the opposite sex were strictly forbidden within apprenticeship indentures − and it was not uncommon at that time for talented male artisans to establish their first solo ventures through dowries. They married that same year and set up home together, living and working in the same building.

Tragedy struck Breguet's life frequently. In the years that followed Abbé Joseph-François Marie, his former tutor and by now one of his closest friends, died in suspicious circumstances. He lost his mother, Suzanne-Marguerite, and stepfather, Joseph, in quick succession, leaving him responsible for his four younger sisters. He and Cécile welcomed a son, Antoine, in 1776, who was the only one of his children to survive into adulthood. Their time together was to be short-lived. Cécile died in 1780. She was twenty-eight. He never remarried. We know little, if anything at all, of Breguet's mental state throughout this time. It was an era when death was closer – it was not uncommon for children and young adults to be orphaned, or for wives to be lost to childbirth – but it is hard to believe that this onslaught of loss had no impact on him or his work. Perhaps watchmaking provided him with an escape and pouring himself into his practice was a coping mechanism. I certainly find that losing myself in the microscopic world of watchmaking is an effective way of disconnecting from reality and shutting off from the outside world.

Yet Breguet's workshop went from strength to strength. Marie Antoinette has been credited as the individual who 'caused Breguet's rapid success and sudden vogue'. She possessed his no. 2 watch,* the second watch he ever made as an independent watchmaker, and recommended him widely both at home and abroad. In 1785 he became Purveyor of Watches to King Louis XVI. Marie Antoinette and Mudge's patron, Queen Charlotte, were friends and exchanged many letters, although they never met face to face. I like to imagine that perhaps they discussed their love of watches, and perhaps Marie Antoinette's recommendation led to the watches Breguet later made for Queen Charlotte and King George III. And of course, with characteristic largesse, Marie Antoinette gave Breguet's watches to those closest to her. She gifted her no. 14 Breguet to Hans Axel von Fersen, a close friend and Swedish count – she had 'AF' in blue enamel added to the case – and commissioned for herself Breguet's no. 46, which, judging from the description, appears to have been a matching watch with the same initials – a little cosy. It's possible

---

* It's described as being self-winding, with day-of-the-month indication, and a repeater.

they were lovers, the watch's *pérpetuelle* symbolising a perpetual love. And some suggest that the anonymous admirer who commissioned the Marie Antoinette watch was von Fersen.

A portrait of Breguet shows a fresh-faced man with an early receding hairline and a look of kindly intelligence. He was clearly never a man for wigs, despite his dealings at court. Accounts of his nature emphasise his modesty and generosity. I particularly appreciate the fact that he was kind to his employees. He is said to have reassured his apprentices, 'Do not be discouraged, or allow failure to dishearten you,' a sentiment that should be framed and mounted above every horological workshop door. He regularly tipped his workers, adding tails to the zeros on their invoices to him to turn them into nines. While tipping is now an act of common courtesy, during the *ancien régime* the idea of volunteering cash to an outworker as a mark of respect would have been virtually unheard of.

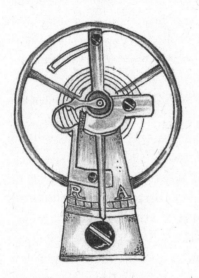

Breguet's *pare-chute* shock setting – he would have been aware of Jean-Pierre Blanchard's ongoing experiments jumping out of a hot-air balloon and using a parachute to soften his fall. Similar to its namesake, the crook-shaped spring on the left of the balance cock allows the staff to bounce very slightly when the watch is knocked or dropped, softening the blow and protecting the pivots from damage.

He sometimes revealed himself to be a showman. In around 1790 he invented the first ever shock setting in a watch. It was a sprung setting that worked to protect the delicate balance pivots from bumps and knocks, which would frequently cause them to break and resulted in a very time-consuming repair. He called it the *pare-chute* (or parachute). To demonstrate how effective his new setting was, legend has it Breguet tested it in front of a crowd of esteemed guests at a party at the home of Charles Maurice de Talleyrand-Périgord, the Bishop of Autun and future First Prince of Benevento. An observer described how Breguet threw 'his watch to the ground, without seeming to harm it in the slightest. "This devil, Breguet," exclaimed the Prince, "is always trying to go one better."' When Breguet retrieved the watch from the floor, guests were astonished to see it was still in perfect working order.

Breguet 'had the power to render even kings the slaves of fashion'. As the century drew to a close, however, that association came to be problematic. In 1792, the French First Republic was born in blood. In the months that followed, the Terror gripped France. Members of the aristocracy, the clergy, and anyone perceived to embody or have an association with the ruling elite of old France were rounded up and imprisoned. By 1794, a series of massacres and mass executions had claimed the lives of thousands of men, women and children. While official figures place the death toll at around 17,000 formal executions, later historians have estimated that the total cost, including those who died in prison or on the run without having been tried, might be as high as 50,000 lives.

Although the most famous method of execution was the guillotine, a grim mascot of the Terror, most of the victims were killed by sword, rifle, pistol or bayonet. Many others died from starvation or disease in France's overcrowded and filthy prisons. Marie Antoinette was detained initially at the Temple prison, where she asked for, and was granted, 'a simple Breguet watch'. Later she was transferred to the Conciergerie in anticipation of her trial. She begged that she would not 'suffer long', but in fact she spent more than two months living in a dank and isolated cell before her trial and eventual execution for treason on 16 October 1793.

The accounts of the survivors who bore witness to the Terror are horrifying. One described how they witnessed a young woman who was forced to drink the blood of a recently executed victim in exchange for her father's liberation. Others tell of fresh blood streaming through pebbled courtyards, or how the condemned would be forced to walk over the dismembered bodies of fellow prisoners to meet their own fate. Executions were a daily public spectacle, attracting huge audiences in street venues like the Place de la Revolution (now the Place de la Concorde). Grisly testimonies made their way, via fleeing survivors, to be recounted in the courts and stately homes of the neighbouring countries who offered them refuge. Europe's ruling classes looked on in horror. The fear that the French Revolution would spread beyond her borders was felt across the continent.

Anyone associated with the king was at risk of losing their life – and Breguet was no exception. A fascinating aspect of Breguet's character is that he had friends in all sorts of places. Despite his cosy relationship with Versailles, he was also very close to political theorist, scientist and French Revolutionary leader Jean-Paul Marat. As the unrest and instability in France intensified even Marat, who had been one of the *ancien régime*'s fiercest critics, fell foul of popular opinion after publishing an attack on the king's finance minister. In April 1793, when an angry lynch mob gathered outside Marat's house to drag him to his fate, it was Breguet who devised a plan to aid his friend's escape. Marat was not aesthetically blessed (he suffered from debilitating dermatitis) and the two made a split-second decision to use his wizened face to their advantage. Using shawls and a dress, Breguet dressed him as an elderly woman and snuck him out of the house on his arm, weaving him through the baying crowds to make their flight to safety. Two months later, when Marat discovered that Breguet himself had been earmarked for the guillotine, he returned the favour by sending him warning and aiding his flight to Switzerland by pulling strings with the powers that be to afford him safe conduct along with what was left of his family: his surviving son Antoine – saving him from conscription into the Revolutionary army – and the sister of his late wife, on the pretext of making his annual business trip to Switzerland. Breguet would never see Marat again. On 13 July 1793 Marat was stabbed in his bathtub by Charlotte Corday, a young Girondin.*

---

* The Girondins were a more moderate faction of the revolutionaries, who also ended up facing mass executions during the Reign of Terror.

Breguet was one of many thousands of people who fled France for nearby countries in this period. And, even for Breguet, life in exile was not easy. Food and materials were all in short supply and refugees were often unwelcome. He would have had to leave most of his tools behind, something that would be immensely painful for any watchmaker, whose tool collections can number in the tens of thousands and feel as personal as fingerprints. Breguet set up a modest workshop, with just a few employees, in Le Locle, not far from his birthplace in Neuchâtel. We can only imagine what the local trade thought of having the most famous watchmaker in the world setting up a competing workshop on their doorstep. He later travelled to London, where for a brief spell he worked for George III. Interestingly, it was during exile that Breguet accomplished some of his greatest technical achievements. Short on tools but still obsessed with his subject, he focused his attention on the invention of new mechanical solutions, including a device known as the *tourbillon*. Breguet conceived the *tourbillon* to overcome something we refer to as positional error. This is the error caused by the effects of gravity on the mechanism as it shifts position with the watch-wearer's movements. This error is most keenly felt by the sensitive oscillating balance, which is also the most vital component for accuracy. Breguet's solution was to house the entire balance and escapement in a constantly rotating carriage, which then evened out the impact of gravitational pull and improved timekeeping accuracy. So important was this new invention that it is labelled on the dial of one his earliest examples, sold to King George IV of Great Britain. The word *tourbillon* was translated into English for the king as the rather delightful 'Whirling-about regulator'.

Nothing was spared in the deconstruction of the *ancien régime*, not even time itself. Time is a loaded construct, filled with associations, whether they be social, political, religious or cultural. Now the Republic rebelled against the authoritarian associations of the Gregorian calendar (the one we still use today), which was seen as a symbolic extension of the establishment. Time under the *ancien régime* had become a symbol of power and control – possessed by the leisured and wealthy at the expense of those less fortunate – so now they embraced a new era by redefining the measurement of time itself. The new Republic was to have a new calendar, with new months, weeks, days, hours and even minutes.

It's something that might seem wildly pointless today – after all, whether there are ten hours in a day or twenty-four, it will make no difference to how many of them we'll live for. But for the Revolutionaries it was an act of starting afresh. This attempt to manipulate time to signify a complete political rebirth has been attempted elsewhere – in Cambodia, the Khmer Rouge declared the year of their takeover 'Year Zero', even though for the rest of the world it was 1975. As Jamaican philosopher Charles Wade Mills phrased it, the resetting of the historical chronometer in the French Republican calendar reflected 'the triumph of reason, light, and equality over the irrationalities and injustice of the *ancien régime*'.

Decimal time, as it was known, literally restarted the clock with the Revolution, redefining September 1792 as 'Year I of the Republic'. This new calendar was still divided into twelve months, but each month was of an equal thirty-day length, with the remaining five days reserved for a series of festivals to mark the end of the year. Each of these thirty-day months was, in turn, divided into a series of three ten-day weeks. The days of the week were renamed, as were the months, to better reflect the seasons. In autumn, *Brumaire*, from the French *brume* meaning 'mist' or 'fog', begins in October and is followed by *Frimaire*, from *frimas* or 'frost', in November. The winter months were *Nivôse*, *Pluviôse* and *Ventôse*, which derived from 'snowy', 'rainy' and 'windy' respectively. Spring began in March, or *Germinal*, the month of germination, then *Floréal* (flowering) and *Prairial* (meadow). As the seasons draw to summer, *Messidor*, from the Latin 'harvest', starts at the end of June, followed by *Thermidor* (hot) and *Fructidor* (fruit) before descending back into autumn with *Vendémiaire*, or 'vintage', in what we would think of as late September.

By tying the seasons to something tangible, the Revolutionaries were liberating time from what they perceived as the oppression of religion and superstition. With many calendrical names linked to Roman gods (such as March to the god of war, Mars, and June to the goddess Juno, wife of Jupiter), the new system eradicated any connection with the gods of old. It was a calendar based on reason, giving time back to the people, to the natural world, and particularly to agriculture. If this is all sounding vaguely familiar, it's because it has striking similarities to event-based timekeeping, albeit with greater emphasis on the structured numerical division of days.

Changing time also changed the face of watches, literally. The new decimal system had further divided the day into just ten hours of an equal 100-minute duration, with each minute being made of 100 seconds, making the decimal second faster than the standard time we operate by today by 0.86 seconds to every decimal. New watches and clocks had to adhere to this new time and examples of the decimal watches made during this brief time survive to this day, including some made by Breguet. They are curious, surreal-looking things. It can take a few moments to realise the reason their dials look odd is because they display ten divisions rather than twelve.

One of Breguet's decimal timekeepers, now in the Frick Collection in New York, not only rises to the challenge of decimal time but also reflects Republican France's eagerness to dispense with myth. The hands of our twelve-hour watches turn in a clockwise direction, replicating our earliest observations of the sun's movement round the Earth as experienced in the northern hemisphere – a movement that since Copernicus (1473–1543) has been known to be based on a misapprehension. For Breguet, a fully metric timekeeper had to embrace rationality and fact, so he designed a dial with two rings – one showing the ten decimal hours, the other a hundred decimal minutes. While the watch hand moves clockwise, every ten decimal minutes the hour ring clicks counterclockwise to reflect the Earth's counterclockwise heliocentric rotation in mechanical form.

In the end, habit proved stronger than ideology, and the life of the decimal clock was short: it was abandoned in less than a year. From its introduction in 1792, the decimal calendar lasted fourteen years until it was ended in 1806 and the Gregorian calendar adopted once more.* Breguet too was only briefly in exile. His cover story about leaving Paris for business reasons only held for so long, especially when the army was looking for his son for conscription. When Breguet's business travel passport expired and he had not returned, he was declared a traitor and a Royalist, and his workshop in Quai de l'Horloge was confiscated and put up for sale. Breguet returned to Paris in April 1795, once the situation had stabilised. Timepieces were in high demand to equip the French army and navy, as well as for scientists, but the city's trade had ground to a near-standstill. The cards were stacked in Breguet's favour. He was a modest man, but he knew his worth, and so he set about not only negotiating the return of his workshop and home at the Quai de l'Horloge, but also demanding that the damage to his business be repaid by the state. Remarkably, with a little help from well-connected friends, the new government returned his house and refitted his workshops at the national expense. An almost unheard-of achievement. The only condition was that he would be up and running within three months. Breguet agreed – on the promise that his workers would be exempt from military service. The deal was done.

Now he extended his business across Europe. One of Breguet's ingenious ideas was that of creating *souscription* (or subscription) watches. For a 25 per cent up-front fee, clients could commission a reliable, everyday, no-frills watch from Breguet; it was still a luxury few could afford, but cheaper than bespoke. The down payment meant Breguet gathered capital up front and could create several similar watches together in a less expensive, serially produced way, which made his watches affordable to a wider range of clients. The scale of production was nowhere near the one we see with our Dutch

---

* Although the litre and metre units of measurement have endured, as did the French franc until recently, which survived until its replacement by the euro on 1 January 2002.

forgeries, but it was a significant move. There are few examples of master watchmakers at any time in history who, once achieving such a high level of celebrity, turned their hand to making their pieces less, not more, expensive. It was an immensely successful business model, with Breguet selling and making around 700 *souscription* watches over the turn of the century.

His chief clientele remained the elite – but a new elite. While in exile he had worked for George III, now he made watches for the bankers and officers of the new Republic; but also for royalty across Europe, including Alexander I, tsar of Russia. Breguet had a bit of a following among the Russian aristocracy and even gets a mention in Alexander Pushkin's *Eugène Onegin*:

> A dandy on the boulevards—
> Strolling at leisure,
> Until his Breguet, ever vigilant
> Reminds him it's midday.

Breguet was as brilliant a diplomat as he was a watchmaker, creating pieces for friends, lovers and sworn enemies alike. Just as he had managed to juggle being *Horloger du Roy* and a friend of Marat, now he made watches for both Napoleon and the Duke of Wellington. Napoleon was so taken with his work he visited Breguet's factory in disguise on several occasions. Arthur Wellesley, 1st Duke of Wellington, owned multiple Breguet watches, including at least one *montre à tact*, which would have enabled him to tell the time by the feel of the case in his pocket. This means it is quite possible that Abraham-Louis Breguet was the unofficial timekeeper of the Battle of Waterloo.

Despite mixing with the very highest members of society, Breguet lived a humble and quiet life. He was described as having a 'young spirit' even in old age. In his last years he was profoundly deaf, but remained cheerful. Breguet's ultimate project, his watch for Marie Antoinette, remained at the core of his focus. There is a note written in August 1832 that confirmed that the queen's watch was still on his bench a month before he died at the age of seventy-six. He had continued to work on it to the end. The piece was ultimately completed by his son and heir, Antoine-Louis.

Just as it's impossible to think of the eighteenth century without considering the French Revolution – a political uprising that sent shockwaves across Europe – so one cannot think about the history of the watch without considering Breguet. His raw talent alongside his political dexterity saved him in an era of continuous flux. In the centuries since, his reputation and his unparalleled watches have endured, with his name appearing in literature as a comment on characters' taste, style and affluence. He is mentioned twice in Alexander Dumas's *The Count of Monte Cristo* and namechecked by Jules Verne and by Thackeray in *Vanity Fair*. Stendhal declared Breguet's watches to be a finer piece of work than the human body: 'Breguet makes a watch that never goes wrong for 20 years, and yet this wretched machine, the body we live with, goes wrong and brings aches and pains at least once a week,' while Victor Hugo went one further, referring, in his 1865 poetry book *Les Chansons des rues et des bois*, to 'God – the mighty Breguet'. To make him a deity might be a stretch too far, but Breguet's watches were certainly a byword for accuracy, utility and divine beauty.

7

Working to the Clock

. . . a stern room, with a deadly statistical clock in it, which measured every
second with a beat like a rap upon a coffin-lid.

Charles Dickens, *Hard Times*, 1854

W hen Craig and I first took our workshop in the Jewellery
Quarter, there was a factory next door. It has since been
knocked down by property developers, who are turning the area
into residential and retail units, but it used to be a huge place:
sprawling old red-brick Victorian buildings that had been clumsily
connected over the years via 1970s office blocks and more modern
aeroplane-hangar-like corrugated-iron warehouses. The factory did
everything from making bus seats to large-scale metal pressing and
forming and precision machining. It was loud – deafeningly loud –
and the constant hum of the motors made our workshop buzz with
white noise. It became such a part of the atmosphere that we stopped
noticing it until the end of the day, when all the machines were
simultaneously switched off and we'd be momentarily stunned by
the silence.

A horn was used to announce the beginning of the day, the start
and end of lunch, and the end of working hours. First thing in the
morning, we'd listen out for the sound of the horn reverberating
around the buildings like the whistle of a steam engine, followed by
the deep, slow whir of motors and heavy machinery firing up. Using
horns to mark time in factories was common through the Industrial

Revolution and beyond, as not much else could cut through the noise of the machines. Today that idea of clock-watching, waiting for the horn, a shared experience with everyone downing tools and walking out the gates together as a place of work grinds to a halt, feels almost quaint. Now that we have mobile phones, remote email access, social media, shift working, virtual meetings and flexitime, there aren't many businesses that just switch off like that. But the advent of clock time in our working lives was as seismic a shift for workers in the industrial era as the late-capitalist, post-pandemic 'wfh' departure from the communal workplace has been in ours. It changed our understanding of how time should be spent.

Factories dominated the skylines of industrialised cities.

Before industrialisation, Britain relied largely on the rhythms of the natural world to dictate the working day and its activities. Working with the land, or the tides, was task-oriented and seasonal. Longer hours could be worked during the extended daylight of the summer months, to make up for those lost to the late dawns and early dusks of winter. For farmers, the harvest was hard work late into the summer evening, which gave way to the cold months when the shorter days could be spent focusing on animal husbandry until things warmed up ready for planting the land again. Crofters – small-scale farmers working on rented plots ('crofts') – turned their hand to building and thatching and, when storms forced them indoors, they made cots or even coffins (quite literally caring for their communities from cradle to grave). Fishermen could mend nets and repair boats when the weather was too poor to catch fish. Life was harsh and unforgiving; but by attending to what was necessary in that moment, this way of working naturally created more labour-intensive seasons as well as those that allowed more time for relaxation and pleasure, albeit with less delineation of the working day. This continues to a greater or lesser extent in agricultural communities today.

The dawn of the predictable, mechanised world of the factory presented a stark contrast, and, from around 1760 through to the Victorian age, put England at the forefront of the Industrial Revolution. This change was less about a switch to wage-labour – even in feudal times, 'dayworkers' had been employed for a fee – than a new stringency about *timed* labour. The working day was no longer defined by sunrise and sunset. As factory processes developed and jobs became more specialised, precise timing became critical for synchronisation. Employees were another cog in the wheel, hired for a duration and budgeted in terms of productivity. Punctuality became profit.

Try to visualise a person of authority any time from the nineteenth to as late as the mid-twentieth century. Perhaps he (it would almost certainly be a 'he') is an industrialist, a factory owner; maybe he's a workshop manager; he could be a politician or trade union leader. Smartly dressed in his dark suit, white shirt buttoned up to the collar. He might be wearing a top hat, a bowler, or even a flat cap. He might have a beard, a moustache, or be clean-shaven. He's almost certainly wearing a waistcoat. If he's wealthier maybe it's silk and patterned, or perhaps it's heavy and woollen, more practical and less ostentatious.* Maybe you're picturing a man like Winston Churchill or Keir Hardie. Perhaps Prince Albert, Abraham Lincoln or even a fictional character like Arthur Conan Doyle's Dr John

---

* The ubiquity of waistcoats, even for men doing heavy manual work in boiling hot jewellery workshops, amazes me.

Watson or Harper Lee's Atticus Finch. Regardless of whether they were new money or old, their social background, or whether they sat on the political right or left, there is something that generally unites the style of these men. Next time you see their photographs or read a description, look out for the chain of a pocket watch pinned to the buttonhole of their waistcoat. (Prince Albert was such a fan of wearing his watch in this way that the chain is now known as an Albert chain.) In this era of industrial expansion, the watch became a symbol not only of its owner's affluence and education, but also of his structured attitude towards work.

A silver pocket watch on an Albert chain, allowing
it to be worn with a waistcoat.

Although Puritanism had disappeared from the mainstream in Europe by the time of the Industrial Revolution, industrialists, too, preached redemption through hard work – lest the Devil find work for idle hands to do. Now, though, the goal was productivity as much as redemption, although the two were often conveniently conflated. To those used to working by the clock, the provincial workers' way of time appeared lazy and disorganised and became increasingly associated with unchristian, slovenly ways. Instead 'time thrift' was promoted as a virtue, and even as a source of health. In 1757, the Irish statesman Edmund Burke argued that it was 'excessive rest and relaxation [that] can be fatal producing melancholy, dejection, despair, and often self-murder' while hard work was 'necessary to health of body and mind'.

Historian E.P. Thompson, in his famous essay 'Time, Work-Discipline and Industrial Capitalism', poetically described the role of the watch in eighteenth-century Britain as 'the small instrument which now regulated the rhythms of industrial life'. It's a description that, as a watchmaker, I particularly enjoy, as I'm often 'regulating' the watches I work on – adjusting the active hairspring length to get the watch running at the right rate – so they can regulate us in our daily lives. For the managerial classes, however, their watches dictated not just their own lives but also those of their employees.

In 1850 James Myles, a factory worker from Dundee, wrote a detailed account of his life working in a spinning mill. James had lived in the countryside before relocating to Dundee with his mother and siblings after his father was sentenced to seven years' transportation to the colonies for murder. James was just seven years old when he managed to get a factory job, a great relief to his mother as the family were already starving. He describes stepping into 'the dust, the din, the work, the hissing and roaring of one person to another'. At a nearby mill the working day ran for seventeen to nineteen hours and mealtimes were almost dispensed with in order to eke the very most out of their workers' productivity, 'Women were employed to boil potatoes and carry them in baskets to the different flats; and the children had to swallow a potato hastily . . . On dinners cooked and eaten as I have described, they had to subsist till half past nine, and frequently ten at night.' In order to get workers to the factory on time, foremen sent men round to wake them up. Myles describes how 'balmy sleep had scarcely closed their urchin eyelids, and steeped their infant souls in blessed forgetfulness, when the thumping of the watchmen's staff on the door would rouse them from repose, and the words "Get up; it's four o'clock," reminded them they were factory children, the unprotected victims of monotonous slavery.'

Human alarm clocks, or 'knocker-uppers', became a common sight in industrial cities.* If you weren't in possession of a clock with an alarm (an expensive complication at the time), you could pay your neighbourhood knocker-upper a small fee to tap on your bedroom windows with a long stick, or even a pea shooter, at the agreed time. Knocker-uppers tried to concentrate as many clients within a short walking distance as they could, but were also careful not to knock too hard in case they woke up their customer's neighbours for free. Their services became more in demand as factories increasingly relied on shift work, expecting people to work irregular hours.

Once in the workplace, access to time was often deliberately restricted and could be manipulated by the employer. By removing all visible clocks other than those controlled by the factory, the only person who knew what time the workers had started and how long they'd been going was the factory master. Shaving time off lunch and designated breaks and extending the working day for a few minutes here and there was easily done. As watches started to become more affordable, those who were able to buy them posed an unwelcome challenge to the factory master's authority.

An account from a mill worker in the mid-nineteenth century describes how: 'We worked as long as we could see in the summer time, and I could not say what hour it was when we stopped. There was nobody but the master and the master's son who had a watch, and we did not know the time. There was one man who had a watch . . . It was taken from him and given into the master's custody because he had told the men the time of day . . .'

James Myles tells a similar story: 'In reality there were no regular hours: masters and managers did with us as they liked. The clocks at factories were often put forward in the morning and back at night, and instead of being instruments for the measurement of time, they were used as cloaks for cheatery and oppression. Though it is known among the hands, all were afraid to speak, and a workman then was afraid to carry a watch, as it was no uncommon event to dismiss

* Knocker-uppers were still going in some towns in the north as late as the 1970s.

anyone who presumed to know too much about the science of horology.'

Time was a form of social control. Making people start work at the crack of dawn, or even earlier, was seen as an effective way to prevent working-class misbehaviour and help them to become productive members of society. As one industrialist explained, 'The necessity of early rising would reduce the poor to a necessity of going to Bed betime; and thereby prevent the Danger of Midnight revels.' And getting the poor used to temporal control couldn't start soon enough. Even children's anarchic sense of the present should be tamed and fitted to schedule. In 1770 English cleric William Temple had advocated that all poor children should be sent from the age of four to workhouses, where they would also receive two hours of schooling a day. He believed that there was:

> considerable use in their being, somehow or other, constantly employed for at least twelve hours a day, whether [these four-year-olds] earn their living or not; for by these means, we hope that the rising generation will be so habituated to constant employment that it would at length prove agreeable and entertaining to them . . .

Because we all know how entertaining most four-year-olds would find ten hours of hard labour followed by another two of schooling. In 1772, in an essay distributed as a pamphlet entitled *A View of Real Grievances*, an anonymous author added that this training in the 'habit of industry' would ensure that, by the time a child was just six or seven, they would be 'habituated, not to say naturalized to Labour and Fatigue.' For those readers with young children looking for further tips, the author offered examples of the work most suited to children of 'their age and strength', chief being agriculture or service at sea. Appropriate tasks to occupy them include digging, ploughing, hedging, chopping wood and carrying heavy things. What could go wrong with giving a six-year-old an axe or sending them off to join the navy?

The watch industry had its own branch of exploitative child labour in the form of what is known as the Christchurch Fusee

Chain Gang. When the Napoleonic Wars caused problems with the supply of fusee chains, most of which came from Switzerland, an entrepreneurial clockmaker from the south coast of England, called Robert Harvey Cox, saw an opportunity. Making fusee chains isn't complicated, but it is exceedingly fiddly. The chains, similar in design to a bicycle chain, are not much thicker than a horse's hair, and are made up of links that are each stamped by hand and then riveted together. To make a section of chain the length of a fingertip requires seventy-five or more individual links and rivets; a complete fusee chain can be the length of your hand. One book on watch-making calls it 'the worst job in the world'. Cox, however, saw it as perfect labour for the little hands of children and, when the Christchurch and Bournemouth Union Workhouse opened in 1764 down the road from him to provide accommodation for the town's poor, he knew where to go looking. At its peak, Cox's factory employed around forty to fifty children, some as young as nine, under the pretext of preventing them from being a financial burden. Their wages, sometimes less than a shilling a week (around £3 today), were paid directly to their workhouse. Days were long and, although they appear to have had some kind of magnification to use, the work could cause headaches and permanent damage to their eyesight. Cox's factory was followed by others, and Christchurch, this otherwise obscure market town on the south coast, would go on to become Britain's leading manufacturer of fusee chains right up until the outbreak of the First World War in 1914.

The damage industrial working attitudes to time caused to poor working communities was very real. The combination of long hours of hard labour, in often dangerous and heavily polluted environ-ments, with disease and malnutrition caused by abject poverty, was toxic. Life expectancy in some of the most intensive manufacturing areas of Britain was incredibly low. An 1841 census of the Black Country parish of Dudley in the West Midlands found that the average was just sixteen years and seven months.

As many of us familiar with the Sunday-night blues can attest, the rhythm of the working week dictates people's perception of time, whether they are working or not. The 1937 Worktown Project, a groundbreaking study of life in the Lancashire town of Bolton, observed that workers would 'anxiously anticipate' the end of time off as much as they looked forward to the end of the working week itself. 'Workers were always looking forward to the end of any prescribed period,' they observed, and could not 'escape . . . from time' even when on their summer holiday.*

In 1954 Philip Larkin railed: 'Why should I let the Toad *work* / squat on my life . . .' Like Larkin, I've never been a happy employee. Unlike factory workers, I have had the liberty to follow my enthusiasms, and have been supported by intelligent and generous people over the course of my career. But there have also been low moments. The truth is, I've always found working for others stressful. Even when the work itself wasn't backbreaking or repetitive, I struggled with the unreadable codes and compromises of the workplace – whether it was the requirement to perform as decorative entertainment to watch collectors at a high-profile auction house or negotiating the hidden social rules of the world's richest people. But more than that, it was the feeling of working *under* someone – the sense of their power over my time, even when they weren't aware of it. Bad management and constantly shifting goals meant that both the labour and the worry of work monopolised my non-working life too. My time felt out of my control, and that brought me to breaking point.

In 2012, I woke up one morning and burst into tears. I was shaking, couldn't breathe, couldn't speak, couldn't move. It felt like someone was squeezing my heart in their fist. It was my first anxiety attack. I knew I couldn't go on. I was an anomaly in the watchmaking world, a square peg that couldn't fit, and the effort was destroying me. I was signed off for stress. Craig, who hated seeing me in

---

* Much of the research was conducted in the pub, where, it was noted, workers drank more quickly on a Friday and Saturday – a trait that was put down not only to the fact that Friday was payday, but also to a desire to stretch leisure time to its utmost.

such a state, said it was time to quit. But he didn't want to see me throw my career away. He had another idea. He had been self-employed before, so perhaps it wasn't such a leap for him to suggest getting a business loan and going off to do our own thing. Being our own bosses. The appeal was instant – the chance to stop handing my day over to someone else; to operate according to my own 'logic of need', as E.P. Thompson put it.

Starting the business was my way of combining work life and home life. This combination isn't always easy, I admit. All small businesses are an extension of their owners, which makes separating them from one's personal life challenging to the point of impossible – particularly when you're working with your significant other. It has been one of the hardest things I've ever done, and on several occasions has nearly cost us everything. But nothing has damaged my health more than subjecting myself to the whims of others. I've learned that, when it comes down to it, I need my time to belong to me.

Controlling time played a fundamental role in making empires possible. We are, even now, living by a Christian timetable, regardless of our religion or lack thereof. The year in which you are reading this has been calculated from the birth of Jesus Christ and defined as AD (standing for *anno domini*, meaning 'in the year of the lord'). Colonialists imposed this Christian concept of time on the people they conquered, seeking to regulate the day just as the first religious orders had once called people to prayer with the toll of the church clock.

The anthropologist Edward T. Hall in the late 1950s coined the term 'chronemics' for the study of the perception of time relating to different cultures. According to Hall, Western nations, especially the United States and northern Europe, are largely 'monochronic' societies, characterised by concentrating on one task and by linear processes. They value punctuality and meeting deadlines, are future-orientated and abhor waiting. They are individualistic. 'Polychronic' cultures, by contrast, such as those in Asia, Latin America, sub-Saharan Africa or the Middle East, tend more to multitasking, are more relationship- than task-focused, present- or even past-oriented (e.g. India, China or Egypt). They may not (as in the Sioux language) actually have a word for waiting. When a monochronic society encounters a polychronic society, the result is often a culture clash. These disparities are even present in the ways we greet each other: while a British or American person might simply say, 'Hi, how are you?' a Mongolian person could spend ten minutes asking about how you slept last night and the health of your family. Globalisation has more recently blurred the differences between national behaviours, with smartphones making fast-tapping, fast-talking multitaskers of us all. But there is no question that in previous centuries these distinctions fuelled racial stereotyping and prejudice.

In the United States, colonisers considered Native Americans to be 'savages'. A major reason for this assessment was that their way of working was still closely entwined with the natural world and they seemed unwilling or unable to embrace the Western system of time. They were 'grounded in their violation of the divine imperative to appropriate the world through mixing their labor with nature'. This happened all over the world. Western observers in the 1800s viewed Mexican miners as 'indolent and child-like people' and noted their 'lack of initiative, inability to save, absences while celebrating too many holidays, willingness to work only three or four days a week if that paid for necessities, insatiable desire for alcohol – all were pointed out as proof of a natural inferiority'. Desire for alcohol aside, in an age where so many of us have completely lost our work–life balance, you could argue that our nineteenth-century Mexican miner was the one who really had his head screwed on. Similar accounts can be found about people from Africa, the Middle East and Protestant-ruled Catholic nations such as Ireland.

This was white, male, European time, designed to favour those who determined the means of dividing it at the expense of those who fell under their control. Placing European time culture at the forefront of our social evolution allowed for the inference that all 'others' were 'behind' in their development. Scholars describe this as temporal othering. In the words of international relations scholar Andrew Hom, it meant that Anglo-European time values were perceived as 'mature, adult, and forward-looking; while other cultures become immature, child-like, and backward'. These kinds of stereotypes went on to underpin Western colonialism in its attempt to reform the world in its own image.

For those who fell under the rule of the new industrial age, any rebellion by the workforce was met with dismissal at best and violence at worst. The overriding message was that those unable to adhere to twelve-hour working days, six days a week, from the age of four, in often extremely dangerous and unimaginably unpleasant conditions, were showing their 'natural inferiority'. Needless to say, the leisure-loving wealthy masters and industrial entrepreneurs were not minded to work in the same way.

Time is a commodity – it's something we possess and can sell. Any job we do is a transaction – we're selling (or maybe leasing) a portion of our time to an employer. If an employer tries to make use of our time without paying for it, we feel rightly cheated. At its most extreme, this arrangement can become enslavement, stripping us of the basic human right of liberty.

In Britain the trade unions, legalised in 1824, understood that time lay at the heart of workers' rights. The first fruits of their struggle came in the Factories Act of 1847, when working hours for women and children were capped at ten hours a day, ceding to demands for the 'three eights' – eight hours' labour, eight hours' recreation, eight hours' rest (a description that makes me think of King Alfred's candle clock).* The next success of the trade union

---

* Holidays with Pay Act of 1938 marked the moment that people could holiday without taking unpaid leave. It wasn't until 1998 that the right to a forty-eight-hour week was enshrined in law.

movement was the Factory Act of 1850, which recommended (it was still ultimately the choice of the manufacturer) that all work should stop at 2 p.m. on Saturday and thus ushered in the modern idea of the weekend.

Traditionally, workers had seized their own time off, taking 'Saint Mondays', an unofficial day of rest taken to sleep off the excesses of Saturday and Sunday night. A hangover (to pardon the pun) from when artisans worked six-day weeks, Monday through to Saturday, Saint Mondays persisted right through to the 1870s and 1880s, despite being much disliked by employers. In 1842 the Early Closing Association, supported by temperance societies, persuaded employers that they would see reduced absenteeism and increased productivity if they allowed workers to finish early on Saturdays. This was promoted as an afternoon for wholesome pleasures and 'rational recreation' – a walk in the countryside, gardening or any other pursuit that required daylight – and boosted a burgeoning leisure industry. Theatres and music halls, which once catered to Monday audiences, began to open on Saturdays, and football clubs, first started by churches to stop workers hitting the alehouse too early, played their matches on Saturday afternoons.*

Railway expansion led to an explosion in the popularity of days out. With the advent of the steam engine radically reducing travel times, day-trippers could get further, quicker, meaning outdoor adventures from picnics and hikes to boating or a trip to the circus were possible. Trips to the coast for health reasons had been popular since Georgian physicians first started touting the holistic benefits of briny waters and fresh salty sea air a century prior, and now there was a peak in Victorian

---

* The Victorian era saw the birth of organised sports. The rules of rugby were published in 1845, the rules of football in 1863. Train travel allowed for local cricket, rugby or football teams to travel for away matches, allowing for larger national competitions, like the FA Cup (1871), to be formed. And it wasn't just the teams that could travel; spectators were able to travel too. Older sporting venues like Epsom enjoyed a boom in visitors as people were now able to travel cross-country to watch the races.

beach-holiday escapades. By the end of the nineteenth century the popular seafronts of cities like Brighton and Blackpool were heaving with tourists from increasingly varied walks of life.

For working-class women, 'time off' never seemed to arrive. The typical ten- to twelve-hour working day was just the beginning for mothers and wives, who were expected to continue their labour when they returned home in the evenings to care for their families. Writing as early as 1739, Mary Collier, a washerwoman from Hampshire, lamented:

> . . . when Home we are come,
> Alas! We find our Work has just begun;
> So many Things for our Attendance call,
> Had we ten Hands, we could employ them all.
> Our Children put to Bed, with greatest Care
> We all Things for your coming Home prepare:
> You sup, and go to Bed without delay,
> And rest yourselves till the ensuing Day;
> While we, alas! But Sleep can have,
> Because our froward Children cry and rave . . .
> In ev'ry Work (we) take our proper Share;
> And from the Time that Harvest doth begin
> Until the Corn be cut and carry'd in,
> Our Toil and Labour's daily so extreme,
> That we've hardly ever *Time to dream.*

A similar eighteenth-century Scottish ballad called 'Answer to Nae Luck about the House' tells a story about a man called John who takes on his wife's chores thinking they'll be easier – only to discover how hard her work actually is. He expresses great relief when his wife at last returns home.

It wasn't simply that women had more work to do, often on top of a day's waged labour (many working-class Victorian women were forced back into work as soon after childbirth as they were able), but that their work in the home didn't adhere to the linear result-oriented pattern of the formal workplace. As most of us know too well, domestic chores – washing, cooking, cleaning – are Sisyphean tasks that are never completed. A meal prepared, eaten and tidied away simply creates space to cook another.

Even now, the work of raising children thwarts our capitalist notion of productivity. New parents working in the home have to adjust to being what the psychotherapist Naomi Stadlen calls 'instantly interruptible', whereby the goal of finishing something, anything, is inevitably thwarted by the more clamorous demands of their offspring. Meanwhile the task of raising a child happens imperceptibly, in an accumulation of mealtimes, bedtimes and cuddles (and later, arguments over screen time). Time passes slowly, at toddler pace, stopping every few steps to look at the ants clustering on a wall at the side of the pavement, and then fast, when we look up to find we suddenly have a hairy teenager.

Why does time seem to speed up as we get older? Although parents witness huge changes in their young children, they themselves remain relatively static.* We tend to remember novelty more sharply, experience it as 'slower' in time. We also, instinctively, use the sharpness of a memory as a guide to its recency. The psychologist Norman Bradbury in 1987 described this as the 'clarity of memory' hypothesis. If a memory is unclear we assume it happened further back, whereas those memories that are magical and life-changing – such as the arrival of a child into our lives – will always seem to have happened barely yesterday.

---

* A recent study by researchers from the Institute for Frontier Areas of Psychology and Mental Health in Freiburg, Germany, and Geneva University has now found that parents do actually perceive time to go faster than non-parents.

As Britain's industrial age progressed, the industry that helped the factories run on time was itself in decline. From the giddy heights of its golden age, watchmaking became one of very few British industries that spectacularly failed to industrialise. The economic damage started early on in the nineteenth century, when the French Revolution and subsequent Napoleonic Wars (1803–1815), combined with competition from Dutch forgeries and the inability of British watchmakers to modernise production methods, dealt a devastating blow to the watch- and clockmaking trades. The British watch industry went from being the centre of the horological world in the eighteenth century to the brink of ruin by 1817.

Thousands of watchmakers were now unemployed and facing destitution. One of the organisers of the Worshipful Company of Clockmakers' relief fund in 1817 described going to visit a former London watchmaker's home, where he found the family in a terrible state with:

> hardly a rag to cover them, and children without shoes or stockings, and in want of bread . . . he had a wife and five children. I found the wife and children in a room without a fire, in the month of January last. Rolled up, in one corner of the room, was something in the shape of a bed on the floor; I believe only a bundle of straw in a cloth without sheets, and a thin sort of cotton covering, which was all the whole seven had to sleep on.

Any trade in luxury goods will suffer during times of war and recession. A luxury watch, as one trader put it to Parliament in 1817, is 'the first article put off in times of distress, and the last put on again when distress is removing'. An 1830s report by the Clockmakers' Company bemoaned cheap and cheerful continental watches, like the Dutch forgeries, being commonly found in jewellers', haberdashers', milliners', dressmakers', perfumers', 'French Toy-shops,' and just 'hawked about the streets.'

Competition led to the declining wages of watchmakers. By the mid-nineteenth century, life for apprentices working in Prescot in

Lancashire was 'mostly hell' and journeymen cutters were known by the unflattering sobriquet of 'poverty knockers'. Unlike Swiss and French watchmakers, the British had been very stubborn about scaling up production to produce more affordable watches. These proud master craftspeople, accustomed to making things of such high quality that they were once regarded as some of the finest toolmakers in the world, had been reluctant to start cutting corners to make lower-quality goods. They had resisted manufactories and even employing women.

That reluctance was not felt in the USA, where a slow start in watchmaking accelerated with the adoption of a mechanised version of Swiss *établissage* techniques. Manufacturers like the Waltham Watch Company, founded in 1850 by the American industrial pioneer Aaron Lufkin Dennison, were able to perfect a standardised, machine-built *ébauches* in very large quantities. They were like the ready-made cake mix of the watch world, almost ready to go, just needing a few final additions before they were ready for the oven.

Historically, even *établissage* watches had a great deal of natural variation, because they were assembled by hand. But in the nineteenth century, American watchmaking really found its footing by combining mass production with standardisation through the use of machines. It meant that by the second half of the century the parts, dials and cases of watches could be made in different locations, allowing company owners to make the most of regional skills and even international differences in metal costs. It also meant that, for the first time, parts could be swapped, so, instead of needing someone like me to make a new balance staff to replace the old one on your watch, a replacement could be ordered from a catalogue of parts. This made them cheaper to put together, cheaper to buy and cheaper to maintain.

In 1896, a New York mail-order business, Ingersoll Watch Company, released the cheapest pocket watch so far, marketed for just $1, the price of a day's wage for the average working American.*

---

* The very cheapest English watches, no more accurate than the Yankee, would have been around $12.50 at the time.

It was called the 'Yankee'. Suddenly people from all walks of life – from servants and factory and rail workers to farmers, cowboys, street traders and even their children – were able to access accurate time whenever they liked. In the twenty years that followed their release, Ingersoll sold forty million Yankee pocket watches, enough to supply well over half the population of the United States at the time. Their slogan was 'The watch that made the dollar famous!'*

Ingersoll watches are technically unremarkable, but are incredible watches nevertheless. They're no-frills pocket watches, in budget nickel-plated cases that feign the appearance of silver, with dials made from printed and pressed paper to imitate white enamel. Their movements are bulky and lack finesse. Some of their parts were stamped for speedy manufacture, leaving rough and rounded edges. They look like they should barely function, but function they did, with great success. Ingersoll marketed their Yankee with a one-year guarantee, promising to repair or replace any watch that failed to keep 'perfect time' free of charge. I've handled some of these watches and, remarkably, they're still just about repairable to this day. They can be dismantled and cleaned; repairs can be made to worn or damaged parts. Budget goods today are designed to be thrown in the bin when they stop working. And yet, this watch that cost just $1 could and still can be serviced just like any other mechanical watch of its era.

If Dutch forgeries had left the British industry walking wounded, the sheer scale and immaculate organisation of American mass production delt the death blow. In 1878 one unnamed 'leading London watchmaker' was quoted as predicting 'that Americans would manufacture common watches for the millions, for this would leave British watchmakers to make aristocratic watches for the hundreds'. It was a prediction that proved all too true – only it failed to anticipate the damage that would wreak on what was left

---

* On a trip to Africa in 1910, 24th President of the United States of America, Theodore Roosevelt, proudly described himself as 'the man from the country where Ingersoll are made'.

of the crippled British industry. In the 1870s and 1880s machinery to create standardised watch movements en masse was shipped to the UK from the USA in a last-ditch attempt to catch up. But it was already too late. By the end of the nineteenth century, the once-thriving community of British watchmakers had dwindled to a few workshops. The last manufacturer to produce watches in Britain on a commercial scale was Smiths, who founded their watchmaking division in 1851 and finally ceased production in 1980. Today there are only a few dozen watchmakers left in the UK who have the knowledge and ability to make timepieces in their entirety using traditional methods. Craig and I are among them. Together, we British watchmakers now produce considerably fewer than 100 watches a year.

8

The Watch of Action

'I refused to take no for an answer.'

Bessie Coleman, aviator, 1920s

Some people just hate to be told something can't be done. I'm one of them. I was the first person in my watchmaking class to tackle a verge watch precisely because my tutor thought it would be too advanced for me. In my final year, instead of a clock I decided to make a pendant watch in the form of a dragonfly. In 2011 I met George Daniels – the world's most famous living watchmaker. He wanted to know why I was working at an auction house and not making watches. Fair question, I thought. He asked whether I wanted to make my own watches some day; I said I did. He laughed loudly, and told me he looked forward to seeing them. It took more than ten years to meet George's challenge. Unfortunately, he's no longer around to see what I'm up to.

This trait goes way back. The first novel I ever read was a response to a similar gauntlet, when an unkind teacher (the type who would lock you in a store cupboard or rip up your work in front of the class because of a spelling mistake) told me I could never read Jules Verne's *Around the World in Eighty Days* because it was too long and hard for me. I was eight, and up to that point had preferred science books to novels, but the story – which was full of daring and adventure – piqued my interest. It's ironic that *Around the World*, published in 1872, is itself the story of a wager: the unbelievable (at that time) idea that it might be possible to circumnavigate the globe in eighty days. Phileas Fogg was prepared to bet (and spend) his fortune defying his doubters as he travelled the world by boat, train, camel and sledge, accompanied by his trusty manservant Passepartout (desperately trying to keep on schedule with the aid of his great-grandfather's watch) and tracked by the doubting Detective Fix.

Phileas Fogg's wager wasn't actually that outlandish. The late-nineteenth-century world was, in some respects, considerably smaller than it had been at the start of the century. Since Richard Trevithick invented the first steam locomotive in 1804, railway fever had spread through Western nations, hurtling people and commodities from one place to another faster than ever before. Railways even shrank the vast nation of America – at least temporally. At the start of the century, it would have taken as long as three months for a single letter to be delivered from, say, New York to New Orleans, and then another three months for the reply to come back. By the 1850s, thanks to the railroad, the whole exchange took just two weeks. Improvements in steam-driven ships and the opening of shipping lanes and canals shortened marine voyages: by 1900, the journey from England to Australia took thirty-five to forty days instead of four months. Meanwhile, the arrival of telegraphs (invented by Samuel Morse in 1844) and telephones (invented by either Antonio Meucci in 1854 or Alexander Graham Bell in 1873 depending on who you believe, as evidence suggests Bell plagiarised parts of his design) allowed people to communicate with friends and family across the world in not months or weeks but moments.

The astonishing aviation advances of the Wright brothers, who first took to the air in 1903, made travel faster still. In 1909 Louis Blériot crossed the English Channel (another wager, with a reward of £1,000 from the *Daily Mail*) in just thirty-six minutes and thirty seconds. It was a period of extraordinary possibility. For all kinds of adventurers (and it seems like there were a lot of them at the time!), once-impossible expeditions now seemed eminently achievable. Men and women travelled to the furthest limits of the Earth, drawn by the mystery of the unknown, the draw of getting there first and the thrill of achieving the impossible. But none of them could have achieved these remarkable feats without watches.

When people travelled the world by foot and horse, the fact that noon shifted as you went east or west barely registered. But as steam trains made crossing a country in a morning possible, the complications became all too apparent – even on a small island like our own. Imagine you've agreed to take the train and meet your cousin, a punctual fellow, at Bristol station on Thursday at two in the afternoon. As you prepare to catch your train in London, you check your watch's accuracy against the town hall clock. Reaching Bristol, you fish it out of your pocket again. It is keeping time nicely: two o'clock exactly. But no sign of your cousin. When he does turn up ten minutes later, he's not flustered and makes no apology. It's not until you see the clock at the Bristol Corn Exchange that you understand: Bristol is actually eleven minutes behind London.* That's because up until the mid-nineteenth century time was regulated by the position of the sun in the sky at noon, when the sun was at its highest.

As travel became faster, it became apparent that these small time differences had big – and potentially dangerous – consequences. Many trains still ran on single tracks, so disparities in an agreed time could lead trains heading in opposite directions straight into each other. To solve the problem, standard national time was introduced on British railways on 1 December 1847, and became law in 1880.

The larger the country, the more extreme the challenges of multiple local times. In a nation as vast as the United States, local time, calculated by the position of the sun, could vary by several hours. Before the adoption of railway time in 1883, it contained over 300

---

* Bristol eventually converted to standard time in 1852, five years after it had been introduced.

Entering service in 1938, *Mallard* broke the world speed record for steam
locomotives the same year after reaching a whopping 126 miles per
hour (203 kilometres per hour) – a record that still stands to this day.

local time zones. Although the new standard time was quick to
catch on in cities, those expected to make the greatest leap from
local to national time were reluctant to make the change: some
rebellious regions refused to adopt national time for many decades.
Dual time observations were common in the US until 1918, and it
was only in 1967 that the 'local option' in time system observance
officially ended.

The next step was a standardised time that could unite the world.
On 22 November 1884, the Greenwich Mean Time system – known
as GMT – was agreed at an international conference in Washington,
DC. It divided the globe into twenty-four zones, each representing
15 degrees of longitude, and an hour of the day. There are a number
of reasons why Greenwich, London, was chosen as the meridian for
this standard. It was, and still is, the site of the Royal Observatory,
which was one of the world's most important locations for the study
of astronomy and horology. It was also chosen because it allows the
date line, on the opposite side of the world to the meridian, to run
through the Pacific Ocean – the only part of the globe that could be

divided without causing residents in the same country to be living across two different days. Travellers now had to start doing what to us is second nature: adjusting their watches to a different country's time zone. It's this that provides the final joke in *Around the World*. Passepartout has consistently refused to change his great-grandfather's watch from London time, insisting that the moon and stars are at fault. He is delighted – and vindicated – when twice on the journey his watch is exactly right. Then, as they arrive back in London, Fogg and Passepartout are crestfallen to discover that their journey has taken eighty-one days. But then they realise that they have travelled eastwards. Thanks to GMT, they have won an extra day and, by extension, the wager.

Jules Verne's story is accurate in another respect: by the late nineteenth century, timekeepers no longer had to be expensive to do as good a job for travellers as the intricate chronometers of the eighteenth century. In 1895, just one year before Ingersoll introduced its famous 'Yankee', Joshua Slocum set off on a journey that would make him the first person to circumnavigate the world single-handedly. He'd chosen to leave his original chronometer at home as it needed repair; he'd been quoted $15 for the job and felt it wasn't worth the expense. Ever the frugal adventurer, he purchased in Yarmouth, Nova Scotia, a cheap alternative timekeeper for the voyage. He described it as his 'famous tin clock, the only timekeeper I carried on the whole voyage. The price of it was a dollar and a half, but on account of the dial being smashed the merchant let me have it for a dollar.' Slocum complained how, 'In our newfangled notions of navigation, it is supposed that a mariner cannot find his way without one [a marine chronometer].' But after three years at sea, and a journey of more than 46,000 miles, Slocum returned triumphant. His one-dollar tin clock had served him loyally throughout.

A quiet village in Nottinghamshire is home to one of the most significant watches in the history of exploration. The Museum of

Timekeeping's collection, housed in Upton Hall, consists of thousands of watches and clocks made over the centuries, donated by all kinds of people. The result is an irregular cabinet of horological curiosities that, to me, offers far more interesting insights into our intimate relationship with our timekeepers than you would find in a more strategically curated collection. Exceptionally valuable and rare chronometers and longcase clocks sit alongside cases of mass-produced 1940s Metamec electric bedside clocks, whose vivid shades of bright orange, retro brown pearlescent and duck-egg blue Lucite acrylic take me back to childhood visits to my grandparents' house.

Housed in a glass cabinet in the museum's watch gallery is an unassuming and rather dishevelled pocket alarm watch that dates back to the early twentieth century. Its dark steel case, the colour of gunmetal, is pockmarked with old reddish-brown rust. The glass that protected its dial is missing, along with the hour, minute and small seconds hand. The only hand that remains indicates that the alarm had been set to chime at about twenty past eleven. Its white enamel dial is still comparatively bright: glass enamel, though brittle and susceptible to cracking if the watch is knocked or dropped, never tarnishes or fades, and keeps its vivid lustre even in extreme conditions. Likewise the black Arabic numerals are as clear as the day they were fired into the enamel. Only the dabs of once-luminous paint that sit on the minute track, marking each hour, have matured from their initial glowing green to a flat dirty brown. Instead of an Albert chain, the bow of the watch is secured to a worn woven bootstring, with a rusty safety pin at the other end. The movement of this watch hasn't run since soon after Thursday, 29 March 1912, the date its owner, Captain Robert Falcon Scott, wrote his last diary entry before he and his remaining crew succumbed to the Antarctic weather, just 12.5 miles from the camp that would have provided them sanctuary.

It's been said that this pocket watch was an essential tool in preventing Captain Scott and his team from sleeping for too long and freezing to death. While I was researching this book, I was fortunate enough to be able to speak to polar explorer Mollie Hughes, who made the trip to the South Pole at a similar time of year to Scott. She said the twenty-four-hour sunlight of the Antarctic summer meant that, once inside the tent, the air was warm enough to sleep in base layers and dry out wet clothes (although her gear was considerably more advanced than Scott's knitted jumpers and gabardines). The greater danger, Hughes found, was the risk of accidentally overexerting yourself by not keeping track of time, because there was no nightfall to indicate the day's end. We use the sun to regulate our days more than we realise. Our biological clocks are geared to wake us up in sunlight and put us to sleep after dark. Without the celestial trigger of darkness, it's harder for our brains to tell us when the day is over and it's time to set up camp. Mollie said the most dangerous point in her trip was not the two-week-long storm she endured at the beginning but when she walked for too long each day trying to make up subsequent lost time. If she was too exhausted to set up camp properly and fell asleep exposed to the harsh Antarctic winds, the consequences would have been fatal.

I scoured Scott's journal, which was recovered along with his watch when his body was found in November 1912, looking for clues that he used his watch to limit his hours of sleep to keep everyone from freezing to death, but he made no mention of it. I did, however, find regular evidence of an almost obsessive regulation of his group's daily routines. His meticulous journal details the time of virtually everything. He marks the time they wake up in the morning, when they start preparing breakfast and how long it takes for all the party to finish eating. Like Mollie Hughes, Scott is mindful of the dangers of disorientation in twenty-four-hour sunlight. He imposes structure on the men's time wherever he can, with a strict regime of travel times and meals, so it's likely he used his pocket alarm watch to chime out when the day's walking had finished and then they would set up camp, ringing the bell for supper, waking the crew in the morning and telling them it was time for breakfast.

He even schedules a service on Sundays, as well as a half-hour slot before it in which that week's hymns were selected, and arranges a series of lectures. It's hard to imagine these explorers – who were, after all, on a life-and-death journey into the unknown – settling down for a pleasant illustrated talk on 'Antarctic Flying Birds', but these activities not only kept spirits high but created a link back to the temporal rhythms of normal life: a sense of normality in the white expanse of nothingness.

When I held Scott's watch in my hands, I felt humbled. It seemed as if his whole life, his hopes, his ambitions, his fears, even the loved ones he left behind, were somehow contained in that otherwise ordinary object. This little machine ventured into the unknown with him and served him until the very end. The mere sight of it conjured up a vivid mental diorama: I imagined the places it had been, the things it had seen, the conversations it had eavesdropped on while discreetly hidden in Scott's pocket. Mechanical watches like this need to be wound to function. With no one around to keep them going, they stop, falling silent with their keeper. They also succumb to the elements: their oil congeals in the cold, their case lets in damp causing ferrous parts to slowly rust, their wheel train gradually seizes solid. To me the museum's decision not to restore this watch back to working order feels like the right one. It would feel somehow disrespectful to Captain Scott to wake his loyal companion from its Antarctic-induced sleep.

As the nineteenth century rolled into the twentieth, watches became indispensable allies of action, and action changed how watches were worn. In the second half of the nineteenth century, soldiers involved in Britain's various overseas campaigns reported the benefits of strapping their fob watches to their wrists: this way they could quickly and easily tell the time without having to fumble around in their pockets in the heat of battle. These wrist-worn watches may have evolved from sweetheart's watches, which young women gave to their lovers as they departed for war. To keep these watches safe, men started making leather pouches, known as wristlets, which held the watch securely while strapped to the wrist. Photographs of British soldiers stationed in northern India show them wearing watches in these 'wristlets' at the time of the Third Burma War in 1885. This was a highly important development: in my view it marks the birth of the commercial mass-market wristwatch as we know it today.

A leather cup strap holding a fob watch,
allowing it to be worn on the wrist.

What started as an improvised trend was quickly capitalised upon by manufacturers. In 1902 Mappin & Webb produced the 'Campaign' watch, an Omega fob watch fitted inside a leather cup wrist-strap. The advertisement promised a 'Small compact watch in absolutely dust and damp-proof oxidised steel case. Reliable timekeeper under the roughest conditions. Complete as illustrated £2.5s. Delivered at the Front. Duty and Postage Free.' The front in this instance was the Second Boer War, fought between Britain and the Boer republics from 1899 to 1902. While Britain's scorched-earth tactics and brutal internment camps have led many to label it our most shameful hour,

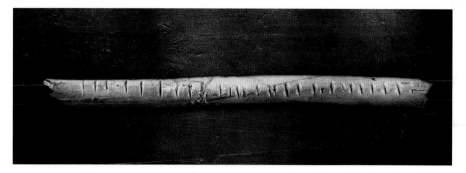

At around 44,000 years old, the Lebombo Bone – carved from a finger-sized piece of baboon fibula – is thought to be the earliest potential timepiece discovered to date. The thirty spaces divided by twenty-nine notches average to a lunar month. Found in South Africa in Border Cave, it is clear evidence of calculation, and the wear patterns indicate this object was regularly used.

The mechanism of a small drum-shaped clock made in Germany some time between 1525 and 1550. The maker's identity is unknown, as it was not unusual at that time for watch- and clockmakers not to sign their work. Made from iron, it was probably created by a locksmith or armourer as the skills required were very similar to those of a watchmaker. These small table clocks are the transitional timepieces that gave rise to the first watches as they were portable and small enough to hold in your hand.

The mechanism is housed in a gilded and engraved drum-shaped canister which measures just under 7 centimetres in diameter and less than 5 centimetres in height. The raised beads that mark every hour were used to tell the time by touch in the dark. It has a single hour hand – this might be, in part, because small clocks and watches of this era weren't accurate enough to warrant one that measured minutes or seconds. It also suggests that more accurate time division wasn't as important to typical owners at the time.

Form watches, so-called as they are, quite literally, made in the form of something, were popular in the mid-seventeenth century. This example – a tiny silver lion that could sit in the palm of your hand – was made in Geneva by watchmaker Jean-Baptiste Duboule in around 1635. To read the time, the lion's belly snaps open and reveals its dial. As Geneva was a Calvinist state in the seventeenth century and banned decorative items like this, the watch would have been intended for export, possibly to the Ottoman Empire.

The mechanism of the watch sits inside the main body of the lion and can be swung out once the lid is opened. The plates are made from gilded and engraved brass. Some of the steel work had been blued (the process of changing the external colour of steel with oxidation caused by heat), a decorative process still used today.

A watch mechanism dating from around 1770, made under the pseudonym 'John Wilter' and proclaiming to have been made in London. Wilter became an obsession of mine after I discovered one of his watches at an auction house. There is no evidence a watchmaker by this name ever existed and the style is not typical of a London-made watch. So-called 'Dutch forgeries' would forever change the dynamic of the watch industry and were the first step in the journey towards making watches affordable to all.

Repoussé watch cases like this one were very popular in the mid- to late eighteenth century. They were stamped or hammered to form a three-dimensional relief which was then engraved, typically with classical or biblical scenes. The technique lent itself to the outer cases of pair-cased watches, with the first of the pair housing the movement and the second protecting the inner.

'Tarts, London' was another notorious fictional watchmaker's name found on eighteenth-century forgeries. This is the front of the case pictured previously. The dial is arcaded, meaning the minute track is scalloped between the hour numerals, which was a popular style among Dutch clockmakers. The watch is fitted to a chatelaine which would have allowed it to be worn on the hip, suspended from a waist band.

An early example of one of Abraham-Louis Breguet's *perpetuelle*, or 'self-winding' watches made in Paris in 1783. Breguet is considered to be among the greatest watchmakers of all time, with a number of his inventions still used in watches today. His *perpetuelle* was the first automatic watch, capable of winding itself using its wearer's movement. The hand and retrograde scale to the top left of the dial show how many hours of wind the mechanism is storing.

The shield-shaped weight swings back and forth as the wearer moves, triggering a series of wheels to wind the spring and power the watch. Breguet also fitted this watch with another of his inventions: steel wires, finely tuned like piano strings, run around the outside of the mechanism and are struck by hammers to chime the hours and quarters. Breguet's wire gongs replaced earlier bells that sat inside the case, allowing repeating watches to become much slimmer, which suited the fashion of the day.

This pocket alarm watch belonged to the explorer Captain Robert Falcon Scott, known as 'Scott of the Antarctic'. It accompanied him on his ill-fated *Terra Nova* expedition (1910–1913) to the South Pole and was found on Scott's body, alongside his companions Edward Adrian Wilson and Henry Robertson Bowers, by a search party eight months after they succumbed to the elements. This watch, along with records and other personal effects, was recovered and returned to the UK.

The Rolex Rebberg, an early mechanism made in the Rebberg manufactory for the Rolex Watch Company in around 1920. They were used in the days before Rolex started making their own movements and are one of our favourite vintage watches to restore. Although mostly machine-made, they were finished and built by hand, creating slight variations between parts that are no longer seen in mass machine-made watches.

One of our favourite, and rarest, early Rolex Rebbergs we've seen was also one of Wilsdorf's first patents: the design of a single-lugged fob watch, allowing it to be suspended from a ribbon or band on a nurse's uniform. It was made in around 1924. The devices on each side are 'vibrating tools' that we use to time in the spring controlling the speed a mechanical watch runs at.

A fingerprint from around 1780 found baked into the counter enamel of a watch dial.
Counter enamel (on the back of the piece) is used to prevent the dial warping when
the visible decorative enamel is applied to the front. This was not designed to be seen,
and it is highly unlikely the person who left the print would have been aware that it
was unique and how special it would be to us today. It was, almost certainly, accidental.
Dial enamelling is a separate art to watchmaking. This watch dial was commissioned
by James Wilson, a watchmaker on King Street, London, whose name is marked
in ink on the back of the dial in the same way that a tailor might put a customer's
name on a bespoke suit to identify it on the rack.

Omega cheerfully reported that its Campaign watch had proved 'life-saving'.*

At home the wristlet was pressed into a different kind of action with the cycling craze of the 1890s, which the *New York Tribune* claimed was more important to humankind 'than all the victories and defeats of Napoleon, with the First and Second Punic Wars . . . thrown in'. In 1893, an advert placed by London retailer Henry Wood claimed that the specialist 'cyclist watch wristlet' was 'the only way to carry a watch on a Cycle without injuring it, and always to hand'. Another, from 1901, describes them as being perfect for 'the tourist, the bicyclist, the soldier'. One of the advantages of marketing these watches for cyclists was that women, who had traditionally worn bracelet watches, were as enthusiastic about the sport as men. That advert features an illustration of a frilly-cuffed feminine hand with a watch round its elegant wrist.

Cyclists weren't the only people of action who needed to see their watches while their hands were occupied. For the world's first pilots, an accurate timekeeper was a life-saving essential, arguably even more so than chronometers were to ships' captains. Pilots didn't simply use them for navigation but to calculate fuel consumption, airspeed and lift capacity. The first dedicated aviation timepiece or 'pilot watch' is said to be the 'Santos', designed by Louis Cartier for the Brazilian aviator Alberto Santos-Dumont in 1904. It came about after Santos complained that he spent far too much time fumbling around in his pocket for his watch when he really needed to be keeping his hands on the plane's controls. Cartier's design had an unusual square case and dial with highly legible black Roman numerals – tough, masculine and perfect for quick reading. The Santos-Dumont watch went into commercial production in 1911 and is still being produced today, more than a century later.

---

* The internment of some 150,000 refugees in British-run concentration camps led to the deaths of over 15,000 native Africans and around 28,000 Boers – three-quarters of whom were children.

Another Swiss watch company, Longines, was, like Cartier, quick to throw its weight behind the wristwatch, as developing a watch suited to the demands of aviation became the longitude challenge of its day. Amelia Earhart wore her Longines chronograph for two of her Atlantic crossings. Another early American aviator, Elinor Smith, set a plethora of records in the late 1920s and 1930s for solo endurance, speed and altitude flights with the assistance of her Longines timekeepers, visible on her wrist in almost every image of her. At sixteen, she was the youngest person in the world to become a certified pilot. Shortly after, she made her name flying under four of New York City's bridges, a challenge she took up after a pilot who had tried the same stunt and failed told her she would never be able to do it. The *New York Times* were so convinced that she too would fail that they prepared her obituary – eighty years before she eventually passed at the ripe old age of ninety-eight. I like to think we'd have got on.

One of the great perils of air navigation was the lack of visible markers to which you could orient yourself: sometimes time was the only marker you had. Philip Van Horn Weems, one of the great pioneers of air navigation, remarked:

> There is no disgrace in being lost in the air. This happens to the best navigators. The important thing is to reduce the periods of being lost or uncertain of position to the lowest limit humanly possible.

It was to Weems that Charles Lindbergh turned when he wanted to learn the art of celestial navigation shortly after completing the first non-stop Atlantic flight, from New York to Paris, in the *Spirit of St Louis* at the age of twenty-five. Together, Lindbergh and Weems invented the Longines 'Hour Angle' watch, with the first calibrated rotating bezel in wristwatch history. The bezel allowed its owner to calculate the angle of the sun relative to the Greenwich meridian: the so-called hour angle that gives the watch its name. Everything about it was adapted to early aviators, from its extra-long strap (for wearing over a bulky flying jacket) to its supersized crown, so that the watch could be wound while wearing flying gloves.

When mountaineer Conrad Anker scaled Mount Everest in May 1999, his goal wasn't to reach the top. Anker was instead trying to work out a mystery. At around 700 metres below the summit and 8,157 metres above sea level, he solved it:

> I was curious, I stopped, turned around, and there was a patch of white. It wasn't snow, it was matt, a light-absorbing colour, like marble. As I got closer, I realised this was the body of one of the pioneering English climbers, frozen onto the mountainside.

The body was male. His right leg was fractured, his arms were outstretched, his clothes had degraded, and the exposed skin on his back was a sun-bleached shade of milky white. Within the remaining tatters of the body's weather-beaten gabardine jacket, Anker found a tag embroidered in crimson thread – 'G. Leigh Mallory'.

Seventy-five years earlier, in 1924, George Mallory and his climbing partner, Andrew Irvine, had gone missing close to the top of Everest as they attempted to become the first to scale the mountain. Until this moment, Mallory's fate had remained one of the great mysteries of mountaineering. To this day, we are still not certain whether he met his end on his journey to, or from, the summit.

Though Mallory's body, which was too difficult to remove, remains exactly where he fell a century ago, a number of his personal possessions were retrieved and are now held by the Royal Geographical Society in London. They include his broken altimeter, snow goggles, knife, matchbox and a silver watch. The watch seems frozen in time. The hands have rusted to dust, leaving nothing but a burnt ochre shadow on the bright white vitreous enamel dial, suggesting the watch last ticked at approximately seven minutes past five, or possibly twenty-five minutes past one, depending on which shadow you read to be which hand. The black outlines of the Arabic numerals still contain a little of the radioactive luminous paint that would have made it easier for Mallory to read his watch at night or in the low light of a snowstorm.

Surprisingly, the watch wasn't found on Mallory's wrist, but in his pocket. There were also no traces of the glass (or crystal) that protected the delicate dial and hands. It's been suggested that it became lost, so Mallory placed the watch in his jacket pocket to protect it from further damage. It's a theory that certainly holds horological weight. In the days before plastic gaskets, which create a friction fit between the crystal and the case of most modern watches, or specialist adhesives that cure under ultraviolet light, watch crystals were commonly fitted using thermal expansion. The bezel, or uppermost band of the case in which the crystal sits, was heated, causing it to expand ever so slightly. The cooler crystal was introduced and, as the bezel cooled and returned to room temperature, it shrank onto the crystal, embracing it firmly. This made the crystal particularly susceptible to extreme changes in temperature. In fact, Mallory's watch has survived in surprisingly good condition for a metal object with ferrous composition that spent seventy-five years near the summit of the world's highest mountain – a testament, perhaps, to the no-nonsense design and robust build quality of these watches.

I made my own journey to Everest – Base Camp rather than summit – in 2011. It was an awe-inspiring experience, of course: how can you not feel insignificant compared to the magnitude of a mountain? Everything took longer because of the altitude, a little like moving through treacle. It gave me an appreciation of why timing is so important for the climbers higher up. Low on oxygen and only able to move very slowly, they often need to be up before dawn in order to pack up camp and make good progress before the ice starts to thaw, increasing the risk of avalanches.

As I made my way up I entered the village of Namche Bazaar, built into the side of the mountain (in Nepal, most things seem to

be built on the side of a mountain), which offered the last experience of something close to civilisation for those headed to the
summit of Everest. I recall seeing streets packed with market stalls
selling sweets, water or hiking gear for those who had damaged or
lost theirs en route. Here what the Sherpas referred to as 'North
Fakes', imported from China, were a reminder of how near we were
to the border. As I looked closer, I started to notice older items, the
kind you might find at an antiques market: a rusty pair of iron cram-
pons, wooden-handled ice axes, like those an explorer might have
used in the 1920s or 1930s. Then I spotted smaller, more personal
items: a pair of spectacles, a wallet. I asked our Sherpas where these
objects came from. They told me that, a few years ago, in response
to the amount of man-made rubbish now cluttering the sides of
Everest, Nepal's government had instructed Sherpas to collect the
scattered detritus. At first they were paid by weight, but when that
fast became too expensive, the government switched to the budget
option of paying by the day. To make back the loss, the Sherpas
started to sell the more interesting objects they found as a lucrative
side income. The original owners of these items were likely dead,
and we would probably never know their stories. I couldn't stop
thinking: what if Mallory's watch had turned up here?

At Chukla Lare, a memorial that climbers pass through as they
approach Everest from Nepal, Buddhist prayer flags and mounds of
stones create a place to pause and reflect on those who have lost
their life on the mountain. There are estimated to be over 100
bodies on Everest, left there as they are too heavy and dangerous
to retrieve. It's accepted by mountaineers that, if they succumb to
the mountain, there's every chance they will be left to become a
part of it.

Mallory's watch, like Scott's before him, would have been a
crucial tool in determining where he was both in time and space.
But you can't take these objects' accuracy for granted. Watches of
this sort have a typical running duration of thirty hours and require
regular winding by hand. Mallory would have had to remember,
every single day without fail, regardless of the weather, his

exhaustion and countless other distractions, to devote some moments to winding his watch. If he faltered, it would falter. If an explorer like Mallory wanted to survive, the first, and last, thing they needed to do was keep their watches alive.

Modern explorers can take their timepieces for granted, as we all do. Whether we use a traditional watch, our computer or a mobile phone, we trust the time to be there whenever we want it. But for those who set out on adventures at the start of the twentieth century, this was not the case. The world might have shrunk but once they had embarked, they were alone and their watch was their only way of ascertaining their whereabouts on this vast, lonely planet. The explorer was reliant on their wits, and their watch.

9

Accelerated Time

They leave their trenches, going over the top,
While time ticks blank and busy on their wrists,
And hope, with furtive eyes and grappling fists,
Flounders in mud.

Siegfried Sassoon, *Attack*, 1918

One day in May 1905 a twenty-six-year-old patent clerk was travelling home from work through central Bern in Switzerland when he heard the town's famous medieval clock – the Zytglogge – chiming out the time. When he looked up at its vast and elaborate clock face he was struck by a curious thought. What would happen if he was sitting in a streetcar travelling away from the Zytglogge at the speed of light? His watch (a Swiss silver pocket watch dating back to about 1900) would carry on ticking out the time as usual, but if he looked back at the clocktower the time would have appeared to stop.

A few months later, the patent clerk – who happened to be called Albert Einstein – published a paper, 'On the Electrodynamics of Moving Bodies', in the German journal *Annalen der Physik*. It would go on to fundamentally alter our understanding of time, the world and the universe. Einstein's theory of relativity argued that time was not an absolute, unchanging ticking clock, as Isaac Newton had claimed centuries earlier, but a flexible dimension that could be stretched and distorted by space, gravity and even personal experience. Einstein demonstrated that time appeared to travel more slowly as gravitational force increased and that it was similarly distorted by the speed of its observer.* Time slows down if you're close to a massive object or if you're travelling at high speed, which has the astonishing result that your clock would run faster at the top of a skyscraper than on the ground floor and slower in a moving car than in a stationary one. While this may not affect the kind of watches I make, it needs to be taken into consideration for GPS systems, which are inside satellites whizzing around at huge heights about the Earth at tremendous speed. For Einstein, time and space were *relative*. 'Time and space,' he summarised, 'are modes by which we think and not conditions in which we live.' He even went so far as to claim that the fundamental concepts of past, present and future are little more than illusions.†

Einstein's revolutionary work on time theory occurred while time-keepers were entering their most rapid state of development. While Einstein was developing his theory of relativity, some 40 miles away on the other side of Lake Neuchâtel, another ambitious young man was developing his own plans. Hans Wilsdorf begun his career as an interpreter and clerk for an exporting firm in the Swiss watchmaking centre of La Chaux-de-Fonds, but in 1903, the twenty-four-year-old German moved to England. He established himself in Hatton Garden, Edwardian London's jewellery centre, just a stone's throw from the capital's old watchmaking centre of Clerkenwell. Wilsdorf

---

* It's worth noting that the subject of relativity itself was not new; it was Einstein's in-depth exploration and rationalisation of it that was revolutionary. Physicists such as Galileo and Lorentz experimented with relativistic mechanics and, much earlier, Polynesian navigators had employed a system whereby their vessel was imagined as a stationary object while the world moved beneath it.

† As Einstein said in a letter to his friend, the engineer Michele Besso, 'For us convinced physicists the distinction between the past, the present, and the future is only an illusion, albeit a persistent one.'

had a plan for a new kind of watch business that was to make him one of the most influential horological entrepreneurs in history. Wilsdorf had read reports of soldiers who had served in the Boer Wars wearing their pocket watches on their wrists. He was convinced that this was the gentleman's watch of the future. Hardly a theory of relativity, you might think – but in 1905 the pocket watch had reigned unchallenged for nearly four centuries, longer even than Newton's theory of gravity.

If an eighteenth-century watch turned up on my workbench, I would genuinely struggle to tell whether it had belonged to a man or a woman. But in the century that followed, the differences became pronounced. As women were increasingly cast as frail and emotionally temperamental, their watches become correspondingly delicate. The pocket watch shrank down to a 'fob watch', worn on a short ornamental chain or pinned like a brooch, while small timepieces mounted on a bracelet or cuff became all the rage. These bangle watches were as much jewellery as they were functioning watches. They were often gold and embellished with vividly coloured enamels, diamonds, split pearls, and gemstones like sapphires, rubies and emeralds. In the nineteenth century, wristwatches were for women.

One of my favourite watches I have ever restored was one such timekeeper. It was made in around 1830 in a lovely warm-coloured gold – a shade you only see in watches and jewellery made before the mid-twentieth century, when higher amounts of copper were used in the alloy. The dial is nestled in a wide bangle formed of gold snakes, decorated with brilliant white and jet-black enamel and gleaming red and green garnet eyes. It reminded me of a warrior's cuff, perfect for Wonder Woman if she ever needed to attend an extravagant nineteenth-century ball, but not the sort of thing you would throw on to pop to the local shop. Practical it was not.

Watches held in wristlets might have been useful for men in action, but civilian men at the turn of the twentieth century considered them effete. Newspaper cartoonists lampooned the new fashion, and men caught wearing them risked being called a sissy. Manly men wore pocket watches – even cowboys. When Levi

Strauss introduced his iconic 501 jeans in 1873, the small inner pocket at their front right was designed to hold a pocket watch – a quirk that persists in jeans to this day. One account from 1900 documents how a trial shipment of wristwatches from Switzerland to the United States was returned by the retailer on the grounds that they were 'unsaleable in the States'. In one 1915 edition of the American comic strip *Mutt and Jeff*, Mutt shows Jeff his new wristwatch. Jeff scoffs: 'Wait a minute – I'll go and get your powder puff.' And yet canny Wilsdorf had a hunch that, with the right marketing, men could be turned on to wristwatches, and that wristwatches would become the defining timepiece of the future.

Wilsdorf didn't have an easy start in life – he was orphaned at the age of twelve and packed off to boarding school in Coburg – but it left him with an independent attitude as well as a small inheritance. He had managed to save a bit from his early employment in La Chaux-de-Fonds, but to get his idea off the ground he needed another investor. His solicitor then introduced him to an Englishman called Alfred Davis. Wilsdorf lost no time in persuading him of the brilliance of his idea: to buy watch movements in volume from a Swiss manufacturer and couple them with pre-made cases to supply the English market.

In 1905 Wilsdorf and Davis began importing movements from the Rebberg district of Bienne, or Biel, in Switzerland from a factory owned by Jean Aegler. These were shipped into the UK and fitted into cases. Some of the cases were made in Switzerland, and some were made by companies like Dennison (founded, as we have seen, by Aaron Lufkin Dennison of Waltham fame), who had a manufactory in Birmingham. The watches were sold under the name Wilsdorf & Davis, with the initials 'W&D' appearing inside the watch cases, while the movements were marked 'Rebberg'. The design was the same utilitarian style chosen by George Mallory for his expedition to the summit of Everest.

Dennison, Wilsdorf & Davis's case manufacturer, had relocated from America to Birmingham in search of a very specific Brummie skillset. Birmingham was the gold and silversmithing capital of the world. At the turn of the century, around 30,000 specialist crafts-people were employed in the trade in and around the Jewellery Quarter. Harnessing their skills meant Dennison was able to establish a factory that became one of the world's most prolific makers of watch cases. By the early twentieth century the Dennison Watch Case Company, located within walking distance of my workshop today, was exporting throughout the USA, and producing cases for Waltham, Elgin and Ingersoll, as well as some of the Swiss greats like Longines, Omega, Jaeger-LeCoultre and now Wilsdorf & Davis.

Ultimately, the Dennison Watch Case Company would succumb to the same fate as the rest of Britain's industry, winding down production before finally closing the factory doors in 1967. A few years ago, Craig and I decided to go and have a look for any last traces of what was once one of the greatest names in the British watch trade. Armed with some 1980s photos of the factory and an old map, we headed to its likely location. All we found was a tarmac expanse of nothing – the building had been replaced by an NHS car park. Deflated, we were about to return home when at the far end of the car park we spotted some grass and an old brick wall. As we approached, a booming voice crackled through unseen speakers. 'You are on CCTV, leave this area immediately! You are on CCTV, leave this area immediately!' By unspoken agreement, we ignored the voice, and Craig gave me a leg-up so I could peer over. The main building had been demolished, but there, next to a crumbling ivy-covered section of factory wall, were the rusting remains of some green industrial rolling mills. They would once have been used to thin down sheets of metal like a rolling pin pressing on a sheet of puff pastry. Behind them I could see a wing of the factory, just one small room and, through the long-since-shattered glass in the square metal window frames, a hint of the dark and overgrown workshop within.

Wilsdorf & Davis had barely got going when storm-clouds started to gather over Europe. The outbreak of the First World War generated a huge level of anti-German sentiment in Britain. The Aliens Restriction Act of 1914 meant Germans in England had to register with police and were forbidden to move more than 5 miles. German businesses were closed down. There were anti-German riots in the streets, and homes were attacked. Manufacturers were keen to avoid any association with the country. The London retailer for Swiss-based manufacturer Stauffer, Son & Co. was forced to issue advertisements to remind the public 'that all the Watch Bracelets supplied by them are British Made as Messrs. S., S. & Co. have never Stocked German Bracelets'. Wilsdorf himself was married to a British woman – Alfred Davis's younger sister – and was a proud Anglophile, but he knew that his conspicuously German name was going to be bad for business. In 1908 Wilsdorf & Davis registered a new name for the company, though it was not until 1915 that they officially started to call themselves the Rolex Watch Company Ltd.

When Craig and I started our workshop we weren't (unlike Wilsdorf and Davis) thinking big. Our small business loan of £15,000 was just enough to secure the rental of our first tiny single-room workshop, buy a couple of old desks (proper watchmaking benches were too expensive, so we built platforms to raise vintage office desks up to the correct height for watchmaking), stock up on some essential hand tools and buy our cleaning and timing machines* (which we still have!). The money ran out almost as soon as it arrived and the first few years were a phenomenal struggle as we figured out how – the irony is not lost on us – to charge for our time, a struggle shared by most creative people. We lived below the poverty threshold for the first eighteen months and regularly ended up selling

---

* A timing machine will listen to the tick of an escapement and tell you, by tracing a line on a graph, whether the mechanism is running fast or slow.

things on eBay to make our rent. We couldn't afford to heat our home in the first winter and ice lined the insides of the walls. On the coldest nights we slept fully clothed in hats and gloves with our cat. Finding the energy to pull sixty- to seventy-hour weeks when you're freezing and living off cheese and bargain pasta is something I never want to experience again. It took seven years before we were on a regular wage.

Craig had developed an infatuation with these very early Rolex Rebbergs at his previous employer, and his reputation for working on them had stuck now we were self-employed, to the extent that clients actively hunted him down. Although early Rolex adverts emphasised the watches' accuracy, announcing them as 'Rolex Watches of Precision' and boasting that they held 'twenty-five world records for accuracy', this was not what drew Craig to them. In fact, in his experience, Rebbergs are not particularly high-quality movements. On close inspection, the edges are rough. Flaws in the design mean they wear themselves out and often need new replacement custom-made components to compensate for their increasingly baggy bearings. The majority have only fifteen jewels in their movements, leaving some bearings subject to unnecessary friction, while the design of the winding stem causes wear to the plates. Sometimes they would have been through the mill with other repairers over the decades and important elements, such as the balance staff, had been replaced with a badly made or adapted part. In early Rolex Rebbergs issues like a later replacement balance wheel that hasn't been set up and adjusted correctly, or the wrong hairspring, both of which cause huge problems with timekeeping, are common. But from the start there was an irresistible aesthetic charm to the Rolex. They are a long way from perfect – but they have an appealing sturdiness to them. Craig describes them as tractors – but they are tractors with an undeniable romance.

Wars – like necessities – are the mothers of invention. Conflict always generates intense periods of investment and innovation in science and technology, because better equipment offers significant advantages on the battlefield. But it also creates unexpected

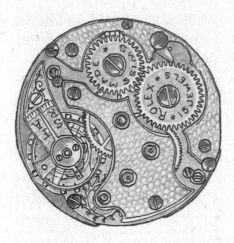

A Rolex Rebberg watch movement dating from around 1920. Craig derived almost as much joy from drawing this as he does from working on them.

inventions, solutions to unforeseen problems. The First World War – which Lenin once called 'the mighty accelerator' – thus gave us blood banks, stainless steel, tanks and drones – but it also gave us the commercial wristwatch.

One thing the war didn't accelerate was Einstein's theory of relativity. The conflict shut down European scientific collaboration, and his theory was only confirmed, by the British scientist Arthur Eddington, in 1919. But in another respect, the war was a perfect fulfilment of his ideas. The conflict was fought on numerous fronts simultaneously, with traditional cycles of time – day and night, even the seasons – ripped to shreds by the perpetual destruction of trench warfare. Technological advances, meanwhile made communication increasingly instantaneous, while the experience of the war created vast gulfs in space – between one side of no-man's-land and the other, between front line and home front – and indeed time, between the prelapsarian pre-war era and the nightmare that followed it. In some ways, the war itself was the theory of relativity made real: a zone in which time and space were destroyed and remade at vertiginous speed.

And yet timekeepers played a vital role in its execution. Fighting on the Western Front was characterised by trench warfare and synchronised attacks. Tactics such as the 'creeping artillery barrage' employed at the Battle of the Somme, which involved sustained artillery fire according to a precise schedule to allow troops to creep closer to the enemy, were highly time-sensitive. As thunderous shelling drowned out orders, timed signals replaced audible calls. Multiple units communicated by telegraph to mobilise at an arranged time. Crawling through the trenches it was near impossible to reach for a pocket watch, so Allied soldiers embraced the wristwatch. Demand was such that it was no longer adapted, as it had been in the Second Boer War, but designed for the purpose. The 'trench watch', as it became known, was fitted with wire lugs to hold the watch directly onto the wrist and could be fitted with a 'shrapnel guard' to protect the fragile glass in battle.

In the Boer War soldiers had had to procure wristwatches for themselves. Now, however, they were retailed in bulk to the military and could be issued to soldiers as kit, along with their uniform, rifle and bayonet, or bought at a discount through Army & Navy stores. Adverts, meanwhile, helped counter any lingering resistance to the wristwatch's historic femininity: 'If HE is fighting at the Front or on the Sea, this Wristlet Waltham will tell him the right time,' proclaimed one in 1914. 'Specially made to withstand rough wear and keep good time under the most trying conditions.' The more affluent officers would often buy fancier versions simply because they could. Gold versions of trench watches are referred to as 'Officers' trench watches'.

Most of the early trench watches were made in Switzerland, where watchmakers had benefited from buying American machine production equipment to boost production. Brands such as International Watch Company (IWC), Omega, Longines and of course Rolex were, unsurprisingly, the first to get in on the action. Their trench watches were simple and functional, with cases formed almost like the sort of pebble you could skim across water and usually made in nickel or brass. These were watches made in volume, created to accompany their owners into the most dangerous of circumstances, whether that be in battle against a human adversary or in Earth's most extreme climates. One 1916 Rolex Rebberg that Craig restored had been bought by the client's grandfather from a man who had served in the Persian Gulf. It came to us tarnished, scratched and dented, missing its bezel and glass. But even with no shock setting or waterproofing, this watch had seen its owner through desert combat and had been worn daily for decades after. It had done what it was designed for.

In the gloom of the trenches at night, soldiers depended on the glow of their watches to read the time. The dials of a trench watch were usually enamel, often bright white, with the numerals applied in luminous radium paint. The hands were skeletonised, pierced out to create hollow spaces down their arm and at their tip, so they too could be filled with glow-in-the-dark paint. The pioneering physicist Marie Curie, together with her husband Pierre, had discovered radium from the uranium-rich and radioactive ore uraninite in 1898. It had rapidly developed a public reputation as a kind of super-element. Thanks to its success in treating cancer, it was promoted as a treatment for everything from hay fever to constipation.* For the watch industry, it was one of radium's decay processes that proved most compelling: when radium was mixed with a phosphor, a radio-luminescent chemical, its action on the phosphor produced a ghostly pale-green fluorescence. The use of radium paint to illuminate the dials of timepieces and scientific instruments quickly caught on. By 1926, US manufacturer Westclox alone was making 1.5 million luminous watches each year. The demand for glow-in-the-dark watch and clock dials, aeroplane instruments, gunsights and ships' compasses was predictably enormous. By the end of 1918, a year after America joined the war, one in six American soldiers owned a luminous watch.

---

* By the early 1900s, factories monetising this new element's glowing properties had sprung up around the world. It was marketed as a health product and incorporated in everything from edibles, like radium-fortified butter and milk, to toothpaste ('for teeth so clean they glow!') and cosmetics. It was even impregnated in clothing and worked into lingerie and jockstraps. The fact that, around the home, radium-laced fly spray was touted for its ability to exterminate pests was not a connection anyone was prepared to make. Radium was a big-dollar business, and those marketing it were keen to suppress any negative press.

Dial factories sprang up across the US and in Switzerland and the UK, with thousands of young women employed to hand-paint the numerals on tens of millions of luminous watch dials.* Radium painting was a prestigious and coveted job – the painters were considered to be skilled artists. Each and every dial went through stringent quality control in a dark room to check its accuracy. Too many mistakes would result in the sack. Out of respect for their talent, wages for radium painters were unusually high, particularly for women of that era. Employers believed that women were the perfect artisans for dial painting, as their small hands were well suited to this intricate and detailed work. It might also be that women were seen as more expendable than men. Male radium

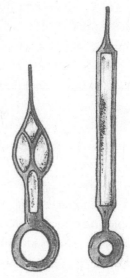

Skeletonised watch hands in-filled with luminous paint.

workers were given protective lead aprons but female radium workers weren't. But women were delighted to take on the work. They were paid by piece, rather than a salary, meaning the more dials they painted each day the more they earned. Some earned as much as three times the average factory worker, taking home more money than their fathers. On average, women could earn the modern equivalent of $370 per week, while the fastest could earn up to $40,000 a year. As many of the factories were established in poor working communities, these wages were life-changing, enabling women to support their families and save for their futures. Although adverts asked for over-eighteens, lax enforcement meant girls much younger managed to gain employment, some as young as eleven.

Radioluminescent paint, given alluring brand names like Undark and Luna, was dubbed 'liquid sunshine'. Dial painters had to mix

* The work and its harrowing consequences in the USA are the subject of Kate Moore's brilliant exposé book, *The Radium Girls*.

their own paint in a crucible using a small dab of powdered radium, a drop of water and gum arabic adhesive. In Switzerland, they applied the paint with glass sticks; in France, they used sticks with cotton wadding on one end; in other countries they used sharpened wood or metal needles; while in the US they used an incredibly fine camel-hair brush to apply the paint in lines as fine as 1 millimetre in thickness. The brushes were so delicate they had a propensity to spread, and so a technique called lip pointing was used, introduced by women who had previously worked painting china ceramics. The 'lip, dip, paint' process involved the painter using their lips to form the brush into a fine point, before dipping it in the radium solution and painting it onto the dial. Although radium paint was equally dangerous wherever it was used, it was this application process that proved so deadly in America.

Initially there were no concerns over its safety – quite the contrary. The dial painters were led to believe that radium was good for them – after all, this was a health product, used in expensive face creams and make-up. Not only were they consuming radium as they formed their brush with their lips, they ended up covered in the radium powder that filled the air of the factories. The workers could be seen glowing with that eerie green light as they headed home in the evening twilight. Local residents remarked that they looked like ghosts. The factories were generally happy places, and the women who worked there felt privileged to do so, playing their part helping soldiers on the front line. Some of the painters were known to play a game of sending secret messages to soldiers by scratching their name and address on the back of the watch case and waiting to see if its eventual owner would write to them. Sometimes they did.

From very early on in its use, management knew there were hidden dangers. Sabin Arnold von Sochocky, the paint's inventor, who eventually died from the long-term effects of radiation exposure, had worked with the Curies, who, by this time, had many radium burns themselves. He himself had amputated the end of his own left index finger after radium became embedded there. While it is true that radium has the power to destroy cancer, it has no ability to differentiate between healthy tissue and a cancerous tumour – it destroys everything in its path. Workshop managers and company executives reassured themselves (comforted no doubt by the huge profits they were making from this booming business) that the quantities of radium the painters consumed were too minute to harm them. But with some girls lip-pointing as much as two times for each numeral, and the fastest workers completing up to 250 dials a day, these amounts added up. The body, mistaking radium for calcium (to which it has a similar chemical nature), delivers it to the bones, where it proceeds to eat away at them – slowly, like a ticking bomb.

This cumulative exposure caused the bones of its victims to rot. Most of the painters' symptoms began with their teeth, which became sore, then loose, then fell out or had to be removed. The gaping holes in their gums rarely healed, causing ulcers and infections that left the bone exposed. As the disease progressed and necrosis of the jaw set in, fragments of bone broke away. The affected women were able to pull chunks of their jawbone out of their mouths.

For those who survived the risks of sepsis and haemorrhaging, the radium was still wreaking havoc in their skeletons, riddling their bones with sponge-like holes until they broke or disintegrated. The pain was excruciating. It left these young women in their teens, twenties and thirties crippled. Cancer was another killer. Rare sarcomas, many starting in the bones, would appear in the years that followed the exposure. Doctors, including those employed by the very companies that manufactured radium dials, repeatedly reassured them that the radium girls' problems were due to feminine nerves, hormones and hysteria – despite tests revealing that these women were now, literally, radioactive. The first fatality, Mollie Maggia from New Jersey, who died in September 1922 at the age of twenty-four, was at first believed to have died from syphilis, on no sounder basis than that she was a young single woman who lived alone.

I once heard about a package of old military watches being stopped by airport security. Decades after their manufacture, they still regis-tered as radioactive. However, the amount of radium in one watch dial is so small it can be disposed of in normal refuse, and many restorers still have drawers and drawers full of movements with old radium-painted dials or 'new old stock' radium-filled hands.* One of Craig's many talents is replicating vintage radium paint with safe modern alternatives. He blends Humbrol model paints to match the colour and uses fine sand or grit to replicate the heavier texture of the original paint, which would fluff up from the surface of the dial like a freshly baked muffin. I once bought a load of old tools at auction and found, lurking among them, a small glass bottle of white dust with the word 'RADIUM' handwritten in ink on an age-browned label, like a sinister version of Alice in Wonderland's 'Drink Me' shrinking potion. I fearfully googled 'how to safely dispose of radioactive material', half expecting some sort of alert to sound and armed police in hazmat suits to burst through the workshop door.

I eventually put the powder in oil to stop it becoming airborne and thought no more of it. The radium girls' lives were blighted. More than fifty had died by the end of the 1920s, though it is impossible to know how many were ultimately affected. Their contribution to the First World War effort was immeasurable, but the price they paid for it was criminal.

Their tragic deaths were not entirely in vain, however. Many of these women, both those who survived and those who didn't, through their consent and that of their families assisted in the research into radiation exposure in the second half of the century. Their bodies helped put a stop to nuclear testing in areas where the fallout could end up in the food chain. The radium girls were the earliest examples of our potential future in an increasingly nuclear world.

---

* 'New old stock' is old stock that's never been used, a bit like things on eBay being sold 'brand new with tags' (only these are 100 years old with tags).

By the end of the First World War it was unusual for a man *not* to own a wristwatch: for many they were a badge of bravery. In the years that followed, trench watches became the foundation of a new era of wristwatch design and innovation. Watch manufacturers created a huge range of new designs and styles including elegant, long, rectangular aerodynamic cases that paid tribute to the emergence of the Art Deco movement. Square watches, whose sides curved outwards (to me they look like a plumped-up sofa cushion), became all the rage too. Where the original trench watches looked much like a small pocket watch with wire lugs soldered on, watches now bore lugs that were integral to the case, like shoulders that extended out and hugged the top of the strap.

Post-war advances in metallurgy and materials science allowed for techniques like gold plating to replace the thicker 'rolled' or 'capped' gold in watch cases. Plating reduced the amount of actual gold needed, bringing the status symbol of a 'gold watch' within reach of a broader group of people. Better materials for base metal cases, like stainless steel, were introduced too, aided by improved equipment that made harder metals easier to machine and finish. Stainless-steel cases had none of chrome and nickel's issues with horrible allergic reactions.

The mechanisms that allowed wristwatches to be automatically wound were also refined, making them more efficient and cheaper to make. They used a weight pivoted at the centre of the mechanism that allowed a wearer to wind the watch simply by moving their wrist. The weight's motion is like that of a wooden rattle swung by a football fan at a match. The spinning weight turns a series of gears, which wind up the mainspring in its barrel, removing the problem of looking at your watch at a critical moment and finding it has stopped.

Complications increased in number and affordability. There were now wrist chronographs, which could function as a stopwatch while maintaining the time, and alarms that could buzz to wake you up in the morning. This steady gain in the popularity of the wristwatch also allowed for profits to be invested into development, improving the quality and accuracy of the movements.

It is this innovative interwar period that both Craig and I were drawn to when we started learning the art of watchmaking – it was one of our early shared enthusiasms. There are so many weird and wonderful designs, some very successful, and many more that were such failures only a masochist would willingly attempt to repair them. I've seen watches from this period that barely functioned when new, let alone after seventy, eighty or a hundred years of wear. Automatic winding, for example, went through many iterations before we arrived at the highly efficient systems we use today. There was 'wig-wag' roller winding, which swings the whole delicate movement up and down its oblong case, and the 'Autorist' with its articulated lugs that flex as the wearer moves their wrist to trigger the winding. We have an Autorist in our collection and, having tried wearing it, I can think of very few publicly appropriate wrist actions vigorous enough to make it reach even half-wind. They're all reminders of our very human urge to test our ingenuity and keep inventing, as well as our need to own the latest exciting technical innovations.

Rolex, more than any other brand, rode this wave of development. In 1919 Wilsdorf and Davis had moved Rolex to Switzerland. After the war ended the British government had imposed heavy import duties on imported watch cases as part of their efforts to fill the treasury again. Rolex kept an office in London until 1931; shortly afterwards the British government abandoned the Gold Standard, gold prices crashed, and Wilsdorf moved to Geneva with the company. Their watches improved vastly in quality over the years, but it was Wilsdorf's genius for marketing that cemented their status. Even the name he had come up with in 1908 – Rolex – had a ring to it. It sounded . . . regal – an association bolstered by model names such as the Prince, Princess, Oyster and, later, Royal, as well as a sister brand called Tudor. Wilsdorf said that the Rolex logo with its five-pointed crown was inspired by the five fingers of a human hand, a nod to the handcraft that goes into every Rolex watch. It still appears on the winder of every Rolex made to this day. These associations have all helped Rolex watches to ooze understated luxury, status and wealth.

Wilsdorf also pounced on publicity opportunities. When in 1927 they developed their first waterproof watch design, the Oyster – so called because the case clamped together like an oyster's shell – they didn't just release it in shops and post a few adverts in newspapers and magazines but sent it across the sea. Mercedes Gleitze wore it round her neck as she became the first British woman to swim the Channel. After her ten hours in the water, the watch was inspected and found to be in perfect working order. Gleitze became Rolex's first ambassador, her famous face promoting the brand, vouching for its integrity and reliability. If today we are used to seeing celebrities and sportspeople advertising watches, Rolex pioneered the practice.

Rolex watches became uniquely identified with the daring achievements of their wearers. In 1933, members of the RAF were the first to fly over Mount Everest, wearing Rolex watches. Rolex printed a series of adverts with the title 'Time flies'. In 1935 an Oyster watch travelled at 272 miles per hour down Daytona Beach with Sir Malcolm Campbell in his legendary car, the Campbell-Railton Bluebird, during one of his many speed record attempts. In the 1950s Rolex launched the Explorer, which was worn by Edmund Hillary and claimed to be the first watch to have reached the summit of the world (although Hillary later stated he was wearing a Smiths watch at the summit). These days you might associate Rolex with Wimbledon, equestrianism, gold at the Masters Tournament, or Formula 1. They sponsor arts festivals around the world. It would be hard to find someone who has never heard of the name Rolex or couldn't identify its iconic crown logo. Rolex, in my mind, were the first watchmakers to make a brand name more prolific than the timekeepers themselves.

The interwar years generated the most rapid evolution of the watch in its 500-year history, a pace of innovation that has continued to the present day. By the time the Second World War broke out in 1939, watches were much better adapted to the adverse and extreme conditions of war than they had been twenty years earlier. Huge pocket-watch-sized wristwatches were supplied to aviators, designed to be read at night with ease and worn over bulky flying gear. Divers' watches accompanied navy frogmen on aquatic missions to raid enemy bases from the sea or plant limpet mines on the outside of an enemy ship's hull. These supercharged wristwatches could be shockproof and water-proof, and their cases could even incorporate a shell that protected them from magnetic fields. Calibrated bezels designed to help early aviators calculate their position mid-air were increasingly used by fighter pilots.

During the First World War, Switzerland's neutrality had been a boon to the watch industry. Swiss watch manufacturers had not lost 50 per cent of their workforce to fighting on the front line, and the country's economy was spared the double blow of reduced productivity and investment in the war effort. Swiss manufacturers didn't simply supply the world's military with completed watches, but also exported kits of movements, cases and dials ready for retail under all sorts of different brand names. In the Second World War, however, trading became more difficult. Switzerland was cut off from trade with the Allies by the Vichy invasion. It was now completely surrounded by Axis powers, posing a quandary for Swiss watch companies, who did not want to deal with Axis powers but were struggling to survive. Early on in the war, Rolex supplied movements to the Italian firm Panerai for use by their Italian navy divers, but it wasn't enough to replace the loss of its main market in Britain.

Eventually, like other watch companies, Rolex started shipping watches to the UK using neutral countries and flagged ships and planes from nations who weren't at war with the UK, such as Spain and Portugal. They were greeted with a built-up demand, particularly among RAF pilots, for whom, since Alex Henshaw became the first person to fly from London to Cape Town wearing a Rolex in 1939, they were practically de rigueur. Wilsdorf went to tremendous lengths to maintain sales with the Allies. He famously offered his watches to British officers who were prisoners of war, to replace those that had been confiscated.* The watches were ordered and then sent via the International Red Cross on the understanding that they would be paid for when the war ended. His gentleman's agreement was a boost for morale, demonstrating that Wilsdorf was confident the POWs would make it through, the war would end and the Allies would win. In one German prisoner camp, Oflag VII-B camp, more than 3,000 watches were ordered by British POWs.

Flight Lieutenant Gerald Imeson, who was interned at Stalag Luft III, 100 miles south-east of Berlin, ordered a Rolex 3525, a top-of-the-range Oyster Chronograph. Imeson used his watch, which had a waterproof oyster-shell case and radium hands and numerals, to light his way in the 334-foot-long tunnel, nicknamed Harry,† which he and his fellow inmates dug as part of a daring escape mission. Imeson was one of 200 men who planned to escape, and had been employed in the run-up as a 'penguin', hiding some of the many tonnes of soil dug up from the tunnels under a bulky coat, in order to redistribute it on the camp grounds.

On the night of the escape, it is possible Imeson used his sophisticated watch to calculate how often the guards patrolled the camp, how long it took each man to crawl through the tunnel, and how many men could enter the tunnel an hour (it worked out at ten). At 1 a.m. the tunnel partially collapsed, slowing the men down. At 4.55 a.m., the seventy-seventh man was noticed by a guard. Those who had made it through ran for their lives. Seventy-three were caught after a manhunt, fifty of whom were executed on the order of Hitler to set an example. Only three managed to get to safety. Imeson was not among them. He returned to the camp and, after being moved

---

* Military-issue watches were often confiscated in POW camps on suspicion that they might contain a compass or tool for escape.

† Harry was one of three tunnels. 'Tom' had been discovered and dynamited; 'Dick' was abandoned when the spot where they planned to surface was built over.

to another POW camp, was liberated in 1945. He treasured his Rolex 3525 for the rest of his life.*

By the end of the Second World War, Rolex was in a strong position but Britain and much of Europe were about to plummet into a recession. The brand focused its attention on the US market, and with some savvy marketing and new designs it managed to compete with home-grown American watch brands like Waltham and Hamilton. Key to their success was their partnership with the world's largest advertising agency at the time, J. Walter Thompson – a relationship that endured for decades.

Thanks to advertising, Wilsdorf's watches were not just time-keepers but storytellers. By associating them with the daring of extreme sports, the precision of extreme timekeeping and the luxury of extreme wealth, the watch proclaimed as much about who you wanted to be as who you were. There is a contemporary marketing term that refers to goods that give the impression of exclusivity and luxury but are intended for the mass market. They are referred to as *masstige* – an amalgamation of the words *mass* (-produced) and *prestige*. This, to my mind, was Hans Wilsdorf's genius.

Hans Wilsdorf and Albert Einstein embarked on very different careers: one was a canny businessman, the other a brilliant theorist. But they had a surprising amount in common. They were born in Germany just two years apart, then both moved to Switzerland, where both of them initially worked as clerks. But for me the most intriguing connection between these two remarkable men was that they both altered our relationship with time. While Einstein overturned centuries of received wisdom about the nature of time itself, Wilsdorf overhauled centuries of assumptions about what a time-keeper could be. We are still living with their legacy today.

* This extraordinary prison break was to become the inspiration for the film *The Great Escape*.

10

Man and Machine

'We are always getting away from the present moment.'
H.G. Wells, *The Time Machine*, 1895

As midday approached on 9 June 1940 the three-man crew of L9323, a light bomber Bristol Blenheim Mk IV aircraft, was returning to base. They had just completed a successful mission bombing a German armoured convoy near Poix-de-Picardie in the Somme area of northern France. The German army was fast advancing on the Allied forces, who were trapped between them and the coast. The Dunkirk evacuations had ended. The crew of L9323 were part of Operation Aerial, whose role was to slow the enemy's progress and give their comrades who had missed the main evacuation as much time as possible to escape across the Channel.

As they passed over Normandy on their return to base, the crew came under enemy fire from a Flak anti-aircraft cannon. Their twenty-five-year-old pilot, Flying Officer Charles Powell Bomford, was killed instantly. The plane's observer, Sergeant Robert Anthony Bowman, pushed his fallen friend's body aside and grabbed the centre stick. Robert had never flown a plane before, but he knew that if he allowed it to freefall to the ground, he and their gunner, Pilot Officer Francis Edward Frayn, would be killed on impact. Wrestling with the controls as the plane came down, he managed to slow their descent enough for them to survive the initial crash. But the impact caused the nose of the plane to push back into the cockpit, pinning Robert against the steering column. He was trapped. Francis rushed to his aid, but was unable to free him. The leaking fuel ignited, and the explosion threw Francis from the plane. Robert died in the fire.

For a long time after, Francis's fate was unknown. Aviation records suggest that he survived and might have been taken as a prisoner of war, though no record of his internment was found. The truth turns out to be far more remarkable.

Francis was stranded – too badly injured to move and aware that the enemy wasn't far away. Lying on the ground where he fell, he heard the marching boots of soldiers approaching. I can only imagine his relief when he heard their voices and realised they had Scottish accents. It was the 51st (Highland) Infantry Division, the last large Allied force left in the area as they too fled to the coast. They rescued Francis, carefully folding his flight jacket beneath his head as a pillow and carrying him by stretcher for two days to the hospital in Saint-Valery-en-Caux from where they all hoped to be evacuated. In the days that followed the 'Miracle of Dunkirk', the last troops – nearly 200,000 Allied personnel and injured soldiers – were boated to safety. Francis was among them. The ship he escaped on was the last to depart before the port fell under enemy fire. It was captained by the esteemed naturalist and British naval officer Sir Peter Scott, the only child of the explorer Robert Falcon Scott.

Francis's rescuers, the 51st (Highland) Infantry Division, were not so lucky. The plan had been to return to collect them, but thick fog made the journey impossible. As dawn arrived on 12 June, the men knew that no further ships would be coming back to rescue them. Trapped, exhausted and depleted, they surrendered that morning.

We know Francis's incredible story because he shared it in detail with his son, who shared it with me, as now I have a tiny part to play in his story too.

At the hospital in France, as Francis's injuries were attended to, a nurse took the battered flying jacket that had been folded under his head and shook it out ready to hang by the side of his bed. There was a metallic *clunk* as a small silvery object hit the floor. She bent down and picked it up. It was Francis's watch, which had somehow also survived the crash. Chunks of metal had been knocked out of the case, the strap had broken, and the rotating bezel he'd used to measure bombing intervals was long gone, but both watch and its owner had made it – battle-scarred but alive and ticking.

And now here it is, sitting before me on my bench. Francis kept the watch after the war and left it to his son, who brought it to me in a Jiffy bag. The dial, once a rich ivory like the colour of full-fat milk, is now patinated with dark speckles. I call it 'foxing' as it reminds me of the dots that appear on the pages of ageing antique books. It's not an official horological term, but one that makes sense to me as someone who loves both old watches and old books. The section of the dial below twelve o'clock bears the remains of what was once the brand name 'MOVADO'.

Movado is a Swiss firm that was founded in La Chaux-de-Fonds in 1905, the same year as Rolex. The founder, nineteen-year-old Achilles Ditesheim, must have been excited by the potential of Esperanto, the artificial international language created by Polish oculist L.L. Zamenhof in 1887, as Movado is Esperanto for 'always in motion'. This model, called a 'Weems' after Lindbergh's navigation expert, Philip Van Horn Weems,* and featuring a movable outer bezel to calculate longitude, was issued to RAF pilots and navigators at the outbreak of hostilities. There were only 2,500 ever made and, looking at it now, I wonder how many are still in existence.

Unusually, the 'MOVADO' on Francis's watch has almost completely, and very carefully, been scratched out – almost as if it was done with the tip of a pin. The only letter remaining visible is the central 'V'. Francis never told his son why the dial had been damaged or what it might mean. Perhaps Francis made the modification himself, to leave the 'V' for Victory. Whatever the reason, it's an important part of the history of

---

* It was in fact a version of the Longines watch that Lieutenant Commander Weems developed with Charles Lindbergh. Longines licensed the design to a number of other companies, Movado included, when they could not keep up with wartime demand.

this little watch, so I will leave it as it is. My primary goal is to get the watch itself back up and running so that Francis's descendants can continue to wear it and remember its story.

Pilot Officer Francis Edward Frayn's Movado Weems after restoration.
With projects like this we're careful to make sure our repairs can
be undone to return the watch to its original state in the future
if desired. The replacement bezel can be removed, and the dents
and chunks knocked out of the case were left untouched.

All watches have stories, but those of the twentieth century feel much closer to us. We learn them directly from their owners, or those related to them, rather than in the pages of books or letters in archives. The watches of the Second World War – whether standard issue or cherished purchase – are imbued with their owners' experiences. Not all of them saw fighting. Some were destined for use by the military's vast number of administrative staff. The marks on military watches, usually printed on their dial and engraved into the back of their case, give us the reference numbers and codes that can tell us which branch they were issued to, in which nation, and in what year. For example, British military watches carry a broad arrow, nicknamed a 'crow's foot' as its three joined lines look a little like a bird's footprint in the sand. There were various codes: *AM* indicated it was issued to the Air Ministry, *ATP* stood for Army Time Piece, and *W.W.W.* meant waterproof wristwatch (the code's literal translation is Watches, Wristlet, Waterproof). The initials R.C.A.F. signify a watch issued to the Royal Canadian Air Force. Other countries had their own systems and this no-nonsense approach to the marking of military-issued pieces usually (there are always exceptions) makes them straightforward to identify and date.

I have also seen watches that were issued to the Nazi military during the 1930s and 1940s, emblazoned with the swastika and eagle of the Kriegsmarine, the *F.L.* Flieger number of a Luftwaffe-issued timepiece, or the *D.H.* property mark that stood for *Deutsches Heer*, meaning 'German army'. On the rare occasions they make their way into our workshop, I swiftly hand them to Craig. His attitude towards them is more clinical than mine. He points out that a Nazi military watch might never have left the stores; or it could have been issued to a junior clerk of little importance, or been traded for some cigarettes by one of the many soldiers who ended up in the Allied prison camps. Equally, any watch that ends up in front of us could, unless its full provenance is known for certain, have been witness to any number of atrocities, which I try not to contemplate. To Craig they're just inanimate objects. You can't blame an inanimate object for the actions of its owner or creator.

Guiltless as an inanimate lump of metal, enamel and glass might be, one can legitimately question the intentions of their collectors. I know owners who see them as nothing more than pieces of history, and whose general interest in twentieth-century military history means they collect a wide range of paraphernalia from both sides. There is, however, another market for Nazi memorabilia that seeks to glorify an appalling phase in our history, and that poses an enduring dilemma for auction houses and dealers. Very recently, a wristwatch claimed to be Hitler's own, a 1933 Huber, came up for sale at auction. Thirty Jewish leaders wrote to the auction house to object to the sale, but it fetched $1.1 million on the first day, and apparently went to a European Jew. A similar discussion raged in 2021, when a watch that had been issued to Chinese soldiers by their government to commend their participation in the June 1989 Tiananmen Square massacre came up for auction in the UK. On the dial, underneath an image of a soldier in a green uniform and helmet, the script reads '89.6 to commemorate the quelling of the rebellion' – a quelling that killed between 300 to 3,000 people, depending on whether you use the official government figures or those of external observers. The auction house initially took the line that this was an 'object of international interest', that its sale wasn't a statement of support, and that the owner had nothing to do with the People's Liberation Army. Yet after the anonymous vendor was subjected to death threats through the auction house's social media and website, the lot was later withdrawn. Objects play an important role in ensuring that the past is never forgotten. But what should we do with artefacts from the darker periods of history? Should they be kept in museums, either on public display or hidden away in a storeroom? Should they be destroyed? Once they reach the open market, we have no way of knowing where they will end up, or how they will be used. There are no easy answers to these questions.

To me, every watch carries the traces of those who have worn it. When the Nazis rounded up Jews for 'resettlement in the east', many believed they were just being relocated. They had little time to pack and limited luggage allowance, so they grabbed their most precious portable possessions. Watches, along with money, clothes, glasses and artificial limbs, were among the first valuables to be confiscated when they arrived at the concentration camps. When the camps were finally liberated, the watches were found heaped in their thousands. Individually, these watches had stories that would never be passed on; collectively they bore witness to humanity's most shameful hour.

The watches gathered after the American bombing of Hiroshima in Japan now form a poignant display at the Hiroshima Peace Memorial Park. When the 9,700-pound (4,600-kilogram) 'Little Boy' atomic warhead landed on the morning of 6 August 1945, 80,000 people were killed outright. The bomb created pressure waves that travelled faster than the speed of sound. Later, it was discovered that each one of the watches caught up in the blast was frozen forever at the time of the detonation: 8.15 a.m.

Through two world wars, wristwatches accompanied men and women through battle and imprisonment, espionage and escape. In the post-war period, watches continued to build on this heroic inheritance. Even after men returned home to civilian life, feats of bravery and endurance were used to sell the watch. The reliability of a watch in all sorts of extreme situations was promoted to the modern man, even if now he was more likely to be called upon to mow the lawn than fight.

Watches sought to outdo each other in chronometry and technical precision. Improvements in water-resistant cases allowed for diving watches to descend to greater depths. In 1960, Rolex strapped its ultimate diving watch, the Deep Sea Special, to the outside of the bathyscaphe submersible *Trieste* and plunged it to a depth of 10,911 metres in the Mariana Trench. It returned to the surface in full working order, which is more than could be said of any human being if they'd gone with it (the current world record for the deepest scuba dive is 'just' 332.35 metres.) In 1965, Omega sent their chronograph the Speedmaster to the moon on the wrists of Buzz Aldrin and Neil Armstrong after it surpassed all competition in its ability to function throughout extreme variations in temperature, changes in pressure, shocks, vibration and acoustic noise that it would encounter on the voyage. Watches for women, by contrast, were more for ornamentation than function, and became more delicate, their dials smaller and smaller. They were almost a reminder that women, after the wartime expansion in their roles, should be back in an apron ready for the most important hour of the day, the return of their husband from work.

Today utility or sports watches are still one of the watch industry's most popular sectors. You might not be able to survive the same conditions as your watch, but at least it will be in one piece to leave to your next of kin.

The science that had led to such devastation in Hiroshima and Nagasaki found a new direction in the period after the Second World

War. It was to change our relationship with our timekeepers forever. As early as the 1930s, Isidor Rabi, a physics professor at Columbia University, had started work on an atomic clock, building on the research of Danish physicist Niels Bohr, who had developed a theory of the structure of the atom.* Bohr had observed that electrons orbit the atomic nucleus with remarkable regularity and that an increase in energy can cause electrons to jump to a higher orbit. As the electrons jump, they emit energy at a specific oscillation frequency. Timekeeping generally depends on things that oscillate, from the pendulum to the balance wheel and hairspring, but the atom's emitted frequency, which Rabi eventually harnessed to produce the first atomic clock in 1945, was found to be more precise and more stable than anything that had come before. It was swiftly followed by further incarnations working with the caesium atom from the National Institute of Standards and Technology (NIST) in Colorado and the National Physical Laboratory in London. By 1967, the General Conference on Weights and Measures had redefined a second as 9,192,631,770 oscillations of the caesium 133 atom. In the years to come, atomic time would make GPS, the internet and space probes possible.

Atomic time was a major breakthrough, but for now it remained locked away in scientific institutes across the world, housed in machines the size of trucks. Meanwhile, other scientific and technological developments were making their way through to civilian time. Electronic clocks, powered by electric impulse rather than swinging pendulum or mainspring, had been around since the 1920s, but now Swiss and American inventors raced to make the technology work for the watch. The first battery-powered wristwatch to make it through the gate was the 1957 Hamilton Ventura, instantly recognisable with its triangular dial and Art Deco-style stepped golden case. Elvis Presley made it covetable by wearing it in the movie *Blue Hawaii*. But the Ventura was launched in haste; its short battery life meant that once sold, many watches were swiftly returned to their retailers, many of

---

* In the 1940s both Bohr and Rabi played a role in the Manhattan Project, contributing to the development of the atomic bomb.

whose repair staff were untrained in the new technology. By the time Hamilton sorted out the Ventura's teething problems, competitors had made up the distance, including the revolutionary Accutron – the name was a selective conflation of *accu*racy and elec*tron*ic – released by the American company Bulova in 1960.

The Accutron kept time with a tuning fork activated by an electronic circuit with a single transistor powered by a small battery. The electronic oscillator assisted the tuning fork to vibrate at a consistent frequency, exactly 360 times a second, which in turn regulated the timekeeping of the watch (Bulova claimed to within +/- two seconds a day), rendering the balance wheel obsolete. To wear, the Accutron both looked and sounded like an object transported from the future. The flagship Accutron Spaceview had no dial, meaning you could see right through to the electronic circuitry of its movement. Pinned to its turquoise-green board were two copper wire coils, supplying the magnetic field to the tuning fork; the rotation of the wheels was accomplished by indexing a tiny wheel with 300 teeth that, in turn, powered a series of gears that drive the rotation of the hands. The tiny tuning fork created a constant audible hum that emanated from the watch surprisingly loudly. (Adverts for the Accutron tried to turn this into a selling point – 'Have you heard the new sound of accuracy? It's the hushed hum of Accutron.') I once slept with an Accutron on my bedside table; it was like sharing a room with a very rowdy bee trapped under a glass. The design of the Accutron Spaceview was a celebration of the very latest in miniature electronics. The transparent design of the dial provided more than a window into the movement; it was a glimpse into the future.

But the watch that decisively overturned the traditional mechanical watch came from Japan. On Christmas Day 1969, Japanese watchmakers Seiko released the Astron, the world's first commercial quartz watch. Instead of the tuning fork, this new invention, the brainchild of Kazunari Sasaki, focused on using piezoelectricity – the process, discovered by Pierre and Jacques Curie in 1880, of using crystals to convert mechanical into electrical energy. When subjected to pressure (the derivation of piezoelectric is from *piezin*, the Greek for 'to squeeze'), a crystal will emit a small electrical pulse which can be used to derive a remarkably stable frequency. This was used to regulate the turning of a magnetic rotor that performed a similar function as an escapement in a mechanical watch. A fashioned piece of quartz can vibrate millions of times a second, compared to 18,000 times an hour in a comparative mechanical watch of the era. The brand-new quartz watch was advertised as being 100 times more accurate than its mechanical rivals.

The Astron was not cheap – only 100 were made initially and sold at Y450,000 (around £10,000 in today's money) – but that didn't remain the case for long. Through massive investment in technology, streamlining production and increasing automation, quartz watch movements became more and more affordable. Today, you can buy a perfectly functioning quartz watch movement for just a few pounds.

It was this speed that took the Swiss by surprise. At the time of the arrival of the Astron, a consortium of Swiss watch companies had, like those in the US, been working on their own versions of the quartz movement for years but, protected by the fixed global exchange rates after the war, the industry itself had failed to innovate and restructure. The Swiss watch industry was still fragmented, with small manufactories, not unlike those that had given it the edge in John Wilter's time, scattered in every town and village of the Jura. Quartz technology required a totally different set of skills – electronics, rather than traditional mechanical engineering – and Japan and Hong Kong were better placed to exploit it than Switzerland and the US.

It's no surprise that the quartz revolution both started and gained pace fastest in the Far East. Japan and Hong Kong had already emerged as world leaders in electronics more generally, with Canon, Panasonic and Mitsubishi becoming hugely successful, and now developed watch companies of their own. Hong Kong had a reputation for producing cheap watches and watch parts for other companies; Japan had brands like Citizen, Seiko and Casio. For the first time in its history, watches were completely machine made, and no longer required skilled artisans to assist in their production. By 1977 Seiko was the largest watch company in the world in terms of revenue.

Meanwhile the watch industry in Switzerland was sleepwalking towards a cliff edge. Swiss watchmakers, just like British watchmakers a century earlier, were too wedded to a belief in mechanical excellence to move with the times. They were slower to invest in new technology and increasingly had to source parts from overseas. This, combined with the rising value of the Swiss franc, priced them out of the low-priced market. By the early 1980s, the Swiss watch industry was in a catastrophic state of decline with mass redundancies and hundreds of companies collapsing, causing recessions in the old watchmaking world.

And if that wasn't enough, hard on the heels of the 'quartz crisis', as it became known in the trade, came another threat: the digital watch.

I can still remember the schoolgirl envy I had for my classmate Victoria's Casio G-shock 'Baby-G'. It was my first year at secondary school. We'd been taken to an Outward Bound centre, supposedly for a bit of team building, and on our first night we were treated to a 'potholing experience'. In reality this was thirty eleven- to twelve-year-old girls crammed into the outer border of the building's loft space, which had been boxed in to create total darkness and filled with various obstacles. We were supposed to circuit the loft in the dark like moles, which would have worked had it not been for the eerie green glow of Victoria's Baby-G digital watch. One touch on the light-up display and we were following it through the darkness. I loved the way it lit up at the push of a button. How I wanted one of my own! But unfortunately it was beyond my frugal parents' means; I would have to wait.

It's incredible to think that the technology that lay behind Victoria's digital watch sprang in part from research undertaken at NASA. The very first digital watch was American, the Hamilton

Pulsar, released in 1972, and used LED technology developed at the Space Agency. It was advertised as 'The ultimate in reliability – no moving parts. No balance wheel, gears, motors, springs, tuning forks, hands, stems or knobs to wind up, run down or wear out!' But the Pulsar, like the Ventura and the Accutron before it, still couldn't compete with the price of products from the Far East. The Japanese Seiko LCD in 1973 and Casio in 1974 left it in the dust.

You might think 'no moving parts' was my idea of hell, but I still genuinely love digital watches. In fact, I have a collection of Casios and they remain my watch of choice for working in the workshop. I suppose I've always been drawn to opposites (I like cats and dogs equally, and my first two albums as a kid were Holst's *The Planets* and *Cyndi Lauper's Greatest Hits*). I may practise and study the traditional craft of watchmaking, relishing the application of skills that are centuries old, but I delight in the plasticky resilience of my Casio. There's something immensely reassuring about wearing a watch that could survive being dropped from the top of a block of flats (or so the adverts say; I haven't tried) when your day job involves applying infinite precision and care to make its high-grade mechanical ances-tors run again.

A late-1970s digital chronograph watch by Seiko.

A Casio doesn't need my love. I don't have to worry about scratching the plastic with sharp metal swarf or bashing it on a milling machine. Nor do I need to repair it: it just goes. If it needs a battery exchange, it's a quick and easy job. And when, one day, it stops and a new battery isn't enough to do the trick, it's not the end of the world: it only cost £30 in the first place.

The salvation of the Swiss watch industry came largely in the form of one man – Nicolas Hayek. A Swiss entrepreneur of Lebanese descent, Hayek was approached by banks to oversee the liquidation of two Swiss watchmaking firms that had been forced out of business by the quartz crisis. But rather than close the businesses altogether, he believed that with substantial restructuring there was a way ahead. He realised that if the Swiss watch industry was going to be saved, it needed to evolve fast, incorporating, rather than rejecting, the competitive quartz technologies, cutting retail prices and presenting something new to the market. So Hayek came up with the idea of producing affordable quartz watches from cheap materials like plastics and resins, in a huge range of bold, fashionable technicolour designs. He called his new brand Swatch.

Swatch quickly cornered the fashion market and single-handedly reinvigorated the analogue watch. In 1985 the *LA Times* called them 'the hottest new fashion accessory on the market'. Swatch watches were so attractive and affordable – Cheryl Chung, then product development manager for Swatch USA, called them 'cheap chic' – that they changed the way people bought and used watches in general. Lanny Mayotte, marketing director for US rival firm Armitron, put it well: 'People today have a wardrobe of watches . . . Years ago you bought a watch for graduation, and it was handed down to the children . . . why not have a fun watch rather than a boring, old, expansion-band watch?'

Five hundred years earlier, a watch had been one of the most expensive personal luxuries money could buy. Now you could purchase one in every colour from your local department store. And if fashions changed? Throw it away and buy a new one. The Swatch watch transformed our relationship with portable time. Yet, ironically, it also saved the mechanical watch. Swatch went on to become such a formidable success that the profits allowed Hayek to buy up faltering historic watch brands and inject new capital into them. The Swatch Group is now one of the largest conglomerates of luxury brands in the world, and owns celebrated marques like Omega, Longines, Tissot and the House of Breguet. The cheap and cheerful watch had rescued the Swiss mechanical watch industry from oblivion.

For American firms and their global outposts there was sadly no such saviour. The strike at US-founded Timex's factory in Dundee, Scotland, in 1993 is among the most notorious incidents caused by the quartz crisis. The picket-line violence has been described as the worst since the miners' strike of 1984. At its peak in the 1970s, Timex was a major employer in the city and the factory employed 7,000 people. By the time it closed, its workforce was reduced to just seventy employees. The trouble arose in 1993 from a dispute over proposed lay-offs, a wage freeze and a reduction in fringe benefits, the result of competition from the Far East. By 1993, Hong Kong was exporting 592 million watches a year and manu-facturers could supply large-quantity batch orders in as little as twenty-five days. Timex in Dundee could not compete. Union workers voted decisively to back strike action rather than face a reduction in workforce or wages. Failing negotiations resulted in the striking workers being locked out of the factory and strike-breakers, or 'scabs' as they were known by the protestors, being bussed in to replace them. One protestor described how: 'Cans of Coke and coffee were thrown over cars and there were incidents of vandalism . . . I always carried a pickaxe handle in the back of my car.' The majority of the Timex workers were women and the strike politicised them. In an interview with the *Scotsman*, one female assembly-line worker said they had changed 'from lambs to lions'. 'There were women that were wee tiny creatures and all of a sudden they were unrecognisable . . . They were fighting for their livelihood. The majority were like me. We had worked there since we were young lassies.'

In the end, after six months of violent unrest, the factory was permanently closed, bringing its forty-seven years as an employer in the city to an end. Though the Timex brand prevailed, and the company still has offices around the world (the bulk of its production is now in Switzerland and the Far East), the closure cast a long shadow. In 2019, when the BBC produced a documentary on 'Scotland's last full-blooded strike', it was clear that the wounds of losing the employer from the city were still raw.

For the watchmaker, one of the greatest changes of the last century was the rapid and at times total shift from craftsperson to machine. The quartz crisis, price warring and budget cuts meant that from the 1970s to the 1990s there was little room left for the skill of the master craftsperson. Humans cost more than machines, and so the more a watch could be made by machine the better. This extended to the way watches were maintained. The Ingersoll Yankee watch, which cost $1, still had the capacity be serviced and repaired. The 1980s saw an increasing number of hermetically sealed watches, whose cases are impossible to open,* meaning the moment they stop working you have little choice but to throw them away and buy another. Most Swatch watches today have a similar construction.

Many watches were no longer expected to last. Plastics, being much softer than metal, wear much faster and infiltrate the delicate mechanisms of watches, further reducing their life expectancy. While a mechanical movement can be repaired by a watchmaker, a circuit board like that found in a battery-operated watch can't: the mechanism is often fused to the case so that when you open it, the components fall apart like a glitter bomb. When one part stops working, the whole watch goes with it. In the decades that followed, built-in obsolescence would become part of the way cars, computers and software were designed.

* I've tried but had to resort to breaking the case.

This was a poignant moment in the long history of our craft. Craig and I are inspired by the era before the quartz crisis, not least because it represents a time when humans and machines worked in unison. Machines improved watch production, efficiency and accuracy but they still needed us to operate them. A 1940s milling machine needs a skilled operator. It requires someone to set it up, monitor it and operate it by hand. It speeds up our work and is beautifully accurate, but it can't be left to its own devices. By contrast, CNC – computer numerical control – will do all the work for you once it's been set up and inputted with a design programme, machining whole components to a near-finished state. It can even be left to run overnight, allowing you to return to your workshop the following morning to find your work has been completed for you while you slept.

It's a surreal thought for a heritage watchmaker. All of our tools and machines are old. Having started our business with a very small loan, around half the price tag of a new Swiss-made lathe, we had no choice but to start buying old equipment that we could restore and tailor for the jobs required. But working with them is one of the most enjoyable aspects of my job. Once you get to know them, you discover that each of our machines has a distinctive personality – so much so that we give them names. Alongside Helga the lathe is her sister, Heidi, another 1950s East German 8-millimetre lathe; they arrived together in a box from Bulgaria. We customised Helga to be a wheel-cutting lathe using the photograph on a cover of a book (ironically called *The Watchmaker and His Lathe*) as reference – she cuts out every tiny tooth on each wheel and pinion within a watch – while Heidi cuts the tiny arbors and stems using a hardened steel graver. There is George the pillar drill, made in the 1960s by the British Ideal Machine and Tool Company (IME), who can drill holes as small as 0.1 millimetres. He sits next to Albert the milling machine, made by Wolf Jahn in around 1900, who works like a drill but whose bed (which holds the work) moves from side to side, allowing him to cut recesses and long trenches into metal. Our smallest lathe, made by Lorch in the 1940s, was found, as one of a pair, in a box of bits on a friend's workshop floor. We named her Maus, or 'mouse' in her native German. Her sister, who we named Spitzmaus ('shrew'), is now an uprighting tool, aligning the tiny holes drilled in different metal plates. These machines need us as much as we need them. We think of them as our colleagues.

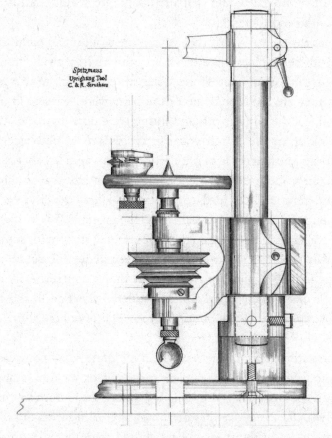

Spitzmaus
Uprighting Tool
C. & R. Struthers

Our uprighting tool, Spitzmaus, adapted for purpose after arriving in a box from our friend's workshop floor.

When Craig and I were training, if I was ever short of a watch to restore for my portfolio he'd let me rifle through his 'tin of movements'. He would beam with glee over his Family Circle biscuit box full of hundreds of old watches and mechanisms, mostly dating from the first half of the twentieth century. While the majority of the other students focused their energy on modern servicing in pursuit of a job at one of the major watch brands' many servicing centres, we'd work on 1920s or 1930s stuff that was often missing parts, cases and case backs, often with no brand name or financial value. Our tutor, Paul Thurlby, a former watch-maker for Omega, would despair: 'Why', he would ask, 'do you only work on old crap?'

For us the appeal is that they offer a trace of human hands. Even though these watches were made at a time when production was becoming more automated, machines were nowhere near as accurate as they are today so some of the parts were very much hand-finished. All the adjustments and fitting were done by hand. When you look at an old watch you can clearly see, if not necessarily errors, the idiosyncrasies of another craftsperson. This always leads me to think about the moment someone first saw that watch in a shop window and decided to buy it, and how precious it then became to them. Did they wear it into the ground? Was it sold back to a jeweller and did it then embark on a new story with someone else? Was it left in a drawer and forgotten about for decades? Or perhaps left to a family member who didn't appreciate its dated style? So many mechanical watches suffered this fate in the 1970s and 1980s when they fell out of fashion. And now here they are, in this Family Circle biscuit tin.

Even earlier trench watches, which are quite small compared to their modern equivalents, regularly met this fate, including those by Wilsdorf & Davis. Their thin wire lugs, often made from soft metals like silver, didn't age well, either physically or in terms of fashion. It's only recently that people have started to realise that they're a century old and worth restoring. We are finally beginning to see the biscuit-tin survivors being repaired and worn again. During the quartz crisis, watchmakers weren't paid much to repair watches and were put under huge pressure to turn work round. We've spoken to now retired watchmakers who've told us that sometimes they were given as little as half an hour to repair a watch. Craig and I can spend a day at the very least, and sometimes several weeks, working on a single mechanism; we're currently working on a restoration that has taken the best part of two years. Back then, the goal was simply to get them ticking as quickly and cheaply as possible. This resulted in a lot of unintended damage. I don't like to hear people knock the work of these watchmakers, who were working in terrible, unsupportive conditions. I've heard of highly skilled watchmakers making more money changing batteries than restoring vintage watches. The quartz crisis was a real low point for traditional watchmaking in so many ways.

Now we are in a new phase again. Technology has overtaken even the quartz watch. The current Apple Watch has more complications than Breguet could have imagined. It's not only a timepiece accurate to within 50 milliseconds, but a phone, internet browser, email provider, car key and fitness tracker, and can even offer ECG and oxygen-level readings – multiple technologies contained in one small package. The moving parts that run the processes in your watch are still quartz, but the time readings they rely on are beamed by satellite and adjusted by atomic clocks. Each time a phone or smartwatch assesses our location, it draws on at least three satellite readings of nanosecond time from space. To tell us where we are, GPS will make adjustments for the relativity between their readings in a way that would have delighted Einstein. Without that adjustment we would register always slightly off, in much the same way that misreading longitude proved fatal for Cloudesley Shovell and his fleet in 1707.

For me, personally, the smartwatch is an advance too far – and somehow invasive. I've never owned a smartwatch. I already feel like my phone and laptop follow me everywhere and that's enough in itself. I love nothing more than getting away from phone signals and Wi-Fi and tracking cookies. I fear that technology will disconnect me completely from the world around me.

When e-books first appeared, everyone claimed the book was

dead. Who needed a book, a bookshop, even a bookshelf, when it could all be accessed on an e-reader? How odd, then, that there was a surprise resurgence in exquisite handmade books, which reminded people of the tactile pleasures of reading. Something similar is happening to watches. Things are starting to come round again. Prices of vintage watches have skyrocketed in recent years. Restoration is valued. Repairs you would have struggled to charge more than a few pounds for in the 1970s are now quoted at tens of thousands by some of the big brands. Today, watches like the Rebberg are increasingly collectable and retailing for good prices, which, in turn, allows for craftspeople to restore them properly. Along with the custom-made larger winding stem to fit their worn plates, we often increase the power of their mainspring to give them every bit of help they can get. Even a tiny difference in the thickness of the spring, maybe just 0.05 millimetres extra, is enough to help improve their accuracy and reliability. There is no spare parts supply, so we hand-make new balance staffs to replace broken ones (another common fault in watches made before shock settings were commonplace), which we turn from steel in our lathe. The delicate pivots that support the entire balance assembly as it oscillates are less than half a millimetre in hight and smaller in thickness. Some of the ones we replace were made quickly and cheaply by one of the quartz crisis repairers, so we make a proper one, which we fit and then 'poise'. Poising is the process of carefully removing minuscule shavings of metal, smaller than a grain of sand, from the balance wheel to make sure its weight is perfectly evenly distributed. It's the miniature equivalent of having your car's tyres balanced.

On a Rebberg, the fine hairspring, made from softer metal than we use in watches today, almost always needs manipulation as, having been through those repair shops, it has usually been bent out of shape. To do this, we use tweezers so fine that their tips are like needles to gently return the hairspring to its perfect spiral. Rebbergs were finished with a human touch. You might have three or four movements that are exactly the same calibre, but you couldn't mix and match the parts. They'd need re-jewelling or modifying, as each movement is coach-built – designed to function only with the parts it has been set up with. Sometimes you'll see that the watchmaker has marked the main components of each movement with hidden dots or scratched numbers to keep them together as they finished and assembled them piece by piece.

In mechanical watches made today, swapping parts is possible with precision CNC engineering. I can't fault CNC: it's an incredible technology and without it many of today's most complicated and accurate watches would be unconceivable. But we'll never embrace it in our workshop. As Craig says, 'I'd rather spend hours messing around with bits that don't fit than have everything finished for me.' That's what it means to restore and rescue something that old; it might take a long time but the process – and the outcome – has soul.

One day there will be watches in deep space. For more than twenty years NASA's Jet Propulsion Laboratory has been developing an atomic clock small enough to take on space exploration missions beyond the reach of GPS.* At present, in order to gain their navigational coordinates, a spacecraft must send a signal from a location to an atomic clock,† and wait for orders – a process that, given the distances involved, can take hours. Atomic clocks need updating several times a day to maintain their phenomenal accuracy. The Deep Space Atomic Clock is currently described as 'about the size of a four-slice toaster', but they are working on making it smaller. It uses mercury-ion trap technology that can maintain a deviation of less than two nanoseconds a day (0.000 000 002 seconds) – accurate enough to enable astronauts in deep space to make navigational

* GPS can guide spacecraft navigationally up to an altitude of approximately 1,860 miles from Earth.
† The clock calculates the position of the spacecraft by measuring the time it takes for electromagnetic waves travelling at the speed of light to travel between the spacecraft and a known location, such as a satellite or antenna.

decisions on their own. Looking for a timepiece small enough and accurate enough that it can guide us as we venture into unknown territories feels like Harrison all over again – the sci-fi sequel.

But Earth time is only relative to Earth, so the further we travel in space, the less relevant it will become. And even though we might now regulate our clocks by the atomic second – the world's most accurate clock (although there is some debate about this!),* at JILA in Colorado (formerly the Joint Institute for Laboratory Astrophysics), is accurate to one second in 15 billion years, roughly the duration of the known universe† – we still live by the circadian rhythm. Habits, as the French Revolution's decimal time showed, are hard to break. But Mars has a similar day duration to ours, so maybe we needn't change just yet. In 2002 a sundial travelled aboard NASA's Mars lander *Surveyor 2001*, and was placed on Mars. Inscribed with the motto 'TWO WORLDS, ONE SUN', it will record the shadows and diurnal rounds of a new planet. It's like starting all over again.

Meanwhile, here on Earth, the internet has transformed our relationship with time once more. Accurate timekeeping now has to be global rather than local, and accurate to within a millionth of a second. Global air travel, telephone networks, banking, broadcasting, all rely on an extraordinary level of time accuracy. Where once we might have asked someone to wait a minute, now we expect everything in a nanosecond, the time it takes light to travel 30 centimetres.

The modern world is terrifyingly fast. I like slow.

The number of beats per second in a modern watch has become a status symbol. The faster a watch ticks, the more accurate it is, as

* The optical lattice clocks at the Observatoire de Paris and the strontium clock at the National Physical Laboratory in Teddington are also strong contenders.
† A standard atomic clock loses a second every 100 million years. They are so accurate that relativity becomes a problem. They can detect the relativistic impact on gravity if they are raised as little as a centimetre.

it becomes less susceptible to changes in position. A sixteenth-century verge watch might tick or beat at most 10,000 times an hour, compared to rates of 18,000 to 28,800 an hour in a modern watch. The fastest escapement in a mechanical watch today can run at a frequency of 129,600 ticks an hour, and at this speed the distinguishable sound of each tick is lost to a consistent hum. Of course, whether it's a slow clunk or high-tech hum, a second or a nanosecond, the pace of the time measured remains the same. But, for me, time can somehow seem more expansive when accompanied by the unhurriedly reassuring *clunk* that verge escapements make every time a tooth from the crown wheel drags itself across a flag and drops onto the next; back and forth, over and over. It's a reassuring sound, like a metronome ticking on top of a piano.

In spacetime, a second might be the difference between landing on Mars and landing tens of thousands of miles away. Here on Earth, I like to remember that the difference in accuracy between the fanciest modern timekeeper and one from the eighteenth century is only a moment – just a few minutes, sometimes seconds, over the course of a day. I can live with that; I've never been one to measure my life by the nanosecond.

II

Eleventh Hour

Though leaves are many, the root is one;
Through all the lying days of my youth
I swayed my leaves and flowers in the sun;
Now I may wither into the truth.

W.B. Yeats, 'The Coming of Wisdom With Time', 1916

I make my living from time – whether it's by making devices that measure it or studying its vast history. It can be overwhelming. Writing this book has been a huge process of bringing together so much that I have learned. Even though I handle timepieces every day of my life, it's only by taking the long view that I've realised just how daunting time can be. It's vast, measured by the movement of distant stars in an infinite universe, and it's also tiny and incredibly intimate, affecting the cells of our bodies right now as you read these words. How we spend our time is personal and also cultural; watches and humans alike are creatures of context.

As a watchmaker first and foremost, I find my place in time through making watches. Leaving work behind in metal feels like my way of leaving a small legacy, something that will live on beyond me, a mechanical ghost to haunt the Earth after I'm gone. Of course, there's no way of knowing how my work will be cared for in the future, if at all. Perhaps, in a few hundred years, the watches we make will sit behind toughened glass in a museum cabinet. Perhaps they will be family heirlooms. Perhaps they'll end up in an old biscuit tin waiting for someone to rescue them. The future is unknown. The past, however, has already unfolded. Considering the history of my subject helps me root myself and my making practice.

One day Craig and I realised that, over the years, we had wound up making virtually every component of a watch in our quest to repair another artisan's work. The time had come to set ourselves the challenge of making every component of a watch of our own. We nicknamed this watch Project 248 as a nod to the way it would be made – by two people, their four hands, and a traditional 8-millimetre watchmaker's lathe. As we surveyed our stable of old tools and repurposed machines, it felt natural to look to the late nineteenth century for our inspiration, to go back to the point in British watchmaking when our home industry was on its last gasp and pick up where our horological ancestors left off.

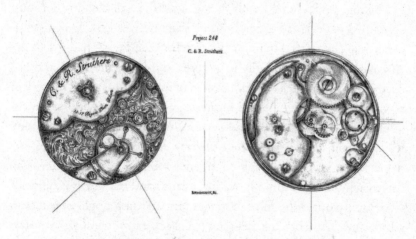

Craig's original concept illustration of Project 248.

We took our design cue from a machine-made fob watch created in the manufactory of a Coventry maker called Thomas Hill in the 1880s. The mechanics and materials of our wristwatch, however, pay tribute to a wider range of watchmakers and companies we admire from across the centuries. It's our little version of al-Jazari's elephant clock, an homage to the international contributions that have made our work possible. As Thomas Hill's pocket watch had no shock protection, we made our own in the style of Abraham-Louis Breguet's 'parachute'. The plates were cut from a metal called German silver, which is a base alloy made of copper, zinc and nickel. The colour, as the name suggests, is silvery in tone with a subtle green-grey hue. It develops a beautiful patina and warmth with age and has been popular among South German watchmakers for over 150 years. The balance, or ticking heart, of the watch was made in the style of one of the greatest watchmakers in living memory, George Daniels, the man who once asked me whether, and when, I would make a watch of my own. Daniels, who lived and worked on the Isle of Man until his death in 2011, was a one-off and an inspiration to Craig and me, not least because as a one-man band he made all the components he needed from scratch, using heritage tools, machines and processes.

Making Project 248 with Craig was the culmination of our combined experience – nearly forty years between us. Ever since I met Craig and started training as a watchmaker, he has had my back. At college he was a vital ally in a workshop where the unusual presence of a girl created a constant source of interest, and rarely in a good way. Now we encourage and motivate each other, something always important in both business and personal relationships, but even more so when you share both. Where we lack confidence in ourselves, it's made up by the unwavering confidence we have in each other. We drive each other on to push our boundaries and test our capabilities.

Bringing our skills and influences together in a single object was a fun, and frequently challenging, mission. From seedling concept to finished object, the whole build took nearly seven years and covered life-changing events for us both. This is how I know that what appears to be a straightforward object can be symbolic of so much more than its form or function. Every element bears the imprint of our hands, our skills and the hours we have taken to work on it. Every part, every stage in the process, is connected with our memories, of moments spanning those seven years.

Of all the tools we might use, none is more incredible than our hands. People often think that because I am a watchmaker I must have very delicate hands. In fact, as everything we do uses fine twee-zers, the physical make-up of your hand means very little. I have watched in awe as a man with hands the size of baseball mitts deli-cately manipulates and restores a balance spring barely thicker than a human hair. What matters is your sensitivity of touch and your understanding of the tolerances of the materials you are working with.

A few years ago, I attended a talk by cardiovascular surgeon Professor Roger Kneebone. I was fascinated by his descriptions of surgery: the years of experience it takes for surgeons to develop the haptic understanding of the body that enables them to predict how the tissues of a child will respond compared to those of an elderly patient, say, or those of a healthy adult compared to an unhealthy one. He described how the vessels in young people are strong and tough, like rubber, whereas when we reach old age they can be as delicate as tissue paper. In watchmaking you learn something simi-lar: brass that is 200 years old will have a different malleability to brass that is new. Steel in a watch made in the sixteenth century will react differently to heat than steel in a watch made in 2020. Ironically, in a watch, older materials, although marked by age and wear, tend to be better quality than new.

Interestingly, a disproportionate number of our clients over the years have been surgeons. In fact, one of the very first customers to entrust us with his precious watches was an orthopaedic surgeon who specialises in hands. Craig and I found it beautiful to listen to him talk about his love of hands, how, 'when opened', the structure beneath the skin is both robotic and alive – almost biomechanoid – operated by tendons, nerves, arteries and bones, with very little soft tissue. Hands are more than objects of human biology, and orthopaedic surgeons work with the physics of our joints. Like watchmakers, they work with saws and files and drills. He says a surgeon should be 'master craftsmen first and foremost',* and

---

* If you had to choose between a surgeon with brilliant theoretical knowledge and one with a track record of dexterity, he would urge you to choose the latter.

describes our hands as 'medical marvels', akin in intricacy to the most complicated of watches.

It is the human hand in 'handmade' that makes a watch unique. When I buff a watch case to achieve a form that is comfortable to wear, I hold a cloth loaded with polishing compound rather than use harder buff sticks or paper. It allows the metal to take its shape according to the curve of my palm, until it is irresistibly tactile. I could hold a machine-made and a handmade watch blindfolded and be able to tell you which is which; until we develop truly sentient artificial intelligence, no machine will be able to replicate the degree of variation you find in a handmade watch.

Handmaking something takes time. When I finish a watch, the last thing I do before it goes in the case is regulate it, adjusting the length of the hairspring to make the time run accurately. In old English watches the index, the pivoted lever we use to adjust the hairspring, is marked 'Fast' and 'Slow' – we included one in the earliest prototype of Project 248.* When I think of making a watch by hand, it is like turning the lever of time to 'Slow'. Indeed, part of the joy of being a craftsperson is that creating something – anything – has its own timeline. Each stage takes as long as it takes, and we have no choice but to give ourselves over to it. Yesterday I spent the whole day filing back the sides of an octagonal case so that it would fit perfectly into an octagonal band. It was only out by a tenth of a millimetre, but it took me nearly eight hours. But now that watch contains the time I have devoted to it. In a fast-moving world, I think there's a generosity to that. Watches not only measure time, they are a manifestation of time – signifiers of the most precious thing we have.

---

* In the end, we gave the 248 a free-sprung mechanism, in which the index is omitted and the hairspring is timed using the mass of the balance. It's a more precise means of regulation.

I have always had a peripheral awareness of the preciousness of time – in a workshop full of watches, how could I not? – but it has taken events in my own life for me to realise it acutely. In June 2017 I woke up one morning with intense pins and needles in my left leg and severe hypersensitivity down the left of my torso. The pain was so severe it reminded me of the time I cracked my ribs when I slipped climbing out of a window as a teenager. I went to the doctor but, as there were no visible marks, he reassured me it was probably just stress. After a few weeks the symptoms stopped. Then, in September that year, I suddenly lost partial sight in my right eye, a symptom accompanied by searing pain. It felt like I'd been punched in the face by a boxer. But again, there were no marks, no swelling. There was nothing noticeably wrong from the outside. This time too, the doctors I first saw tried to pass it off as stress, but I was becoming convinced there was something very wrong, and so fought for further tests. A few months later, after multiple referrals and an MRI scan, I was diagnosed with multiple sclerosis.

I will never forget those months undergoing tests. No one had taken the time to explain the possibilities to me, and so I feared it was a brain tumour, the only diagnosis I could rationalise out of my weird neurological snags. I'd never thought I was immortal, but the sudden intimation that I might be facing death a lot sooner than I'd anticipated changed the way I experienced the world. I remember that winter there was a sudden heavy snowstorm. The house we lived in backed onto a park, and when the snow settled a hill near the foot of our garden provided endless entertainment for children out sledging. I insisted on dragging Craig, who hates the cold, out on a long lap of the park, detouring from what would have been a straightforward trip to the supermarket at the end of our road, because it was rare to see snow that heavy in Birmingham and I wanted to experience it while I had the chance. That memory – of the ice-cold breeze biting my face, the way the light bounced off the white ground and trees, the sounds of children playing and excited dogs barking muffled against the soft snow, is as vivid today as that day years ago. I wanted to soak up the world while I still could.

It wasn't death that scared me so much as how I had spent my time. I had worked and worked – for what? What did I have to show for it? I had spent years being stressed, anxious and exhausted, but I hadn't allowed for happiness, and now I might be out of time.

As it turned out, I was lucky: life expectancy for people with multiple sclerosis has improved in recent decades. It's not pleasant, walking around with what feels like an undetonated Second World War bomb in your head, which may sleep peacefully for the remainder of my life, or go off at any moment, but things are far more promising than they used to be. I am incredibly fortunate to be living at a time when treatment options for MS have improved. I stand a good chance of living what I refer to with my neurologist as a 'long and blissfully uneventful life'. I'm also glad to live in a country that affords me access to one of the finest publicly funded healthcare systems in the world. I have a lot to be grateful for, and I just wish it hadn't taken preparing myself for a much worse diagnosis to realise that.

The French composer Louis-Hector Berlioz (1803–1869) famously declared: 'Time is a great teacher. Unfortunately, it kills all its pupils.' We are all students of time. Certainly, the lesson I've learned is how precious time really is.

My diagnosis has permanently changed how I want to live these seconds, hours, days I've been given. Those first doctors were right in one respect: a lot of my condition can be blamed on stress. Every single one of my relapses has followed shortly after an anxiety attack, and I've not had an anxiety attack that didn't precede a relapse. I used to see stress as a badge of honour, an internal scar I wore that proved how brave I'd been, throwing myself into high-pressure situations, surviving bullying and discrimination, and coming out still fighting. I operated at a fever-pitch of anxiety. Now I realise tolerating stress is not something to be celebrated. It takes just as much strength and resilience to say no, and to know when to walk away. Knowing how ill stress can make me, I avoid it like the plague. I'm a reformed workaholic.

Sadly, I can't claim that I now lead a flawlessly calm life, at peace with the universe. I still get wound up over stupid things, like energy companies failing to manage my account properly or my painfully slow-loading computer. But I am far more relaxed about life than I have been in the past. I focus my energy on the people in my life who support me; those who don't aren't worth my time. I allocate more time for taking in and appreciating the world around me.

The truth is, of course, all of us have limited time, and the amount we have is not within any of our control. Yet how we spend it is, and our experiences can alter our perception of time beyond the minutes logged on our watches. That day when Craig and I walked in the snow is lodged in my mind as if it were yesterday, defying clock time. Recent studies in neuroscience support the theory that our experience of time is tethered to the quality of our experiences.

Philosophers have known for centuries that there is a difference between having lived for a long amount of time and having lived an eventful, active and rich life. Old age isn't always a sign of experience or wisdom. There's a wonderful quote in *On the Shortness of Life*, written in around 49 AD by the Stoic philosopher Seneca, where he sums up the contrast between time lived and experience gained through the metaphor of a storm:

> you must not think a man has lived long because he has white hair and wrinkles: he has not lived long, just existed long. For suppose you should think that a man had had a long voyage who had been caught in a raging storm as he left the harbour, and carried hither and thither and driven round and round in a circle by the rage of opposing winds? He did not have a long voyage, just a long tossing about.

The experience of living a full life might leave us feeling as if our lives are flying past us, but making vivid memories means that in years to come we will look back on a life that feels long.

We all measure our lives in moments of time, and the memories that accompany them. Watches, which tell the time for us as they did for our relatives before us, provide a constant in those memories. People who would not consider themselves 'into watches' still want the treasured pocket watch they've inherited from their great-grandfather restored, even if they know they'll never use it. Looking at the dial of an old watch, we see the same hands passing the hours and minutes that our parents, grandparents or great-grandparents saw, hear the same ticking sound that they heard as they measured the passing moments of their lives. And then, if we are lucky, we wind it up and carry it with us as we go through ours.

Fast and slow, an English regulator.

How to Repair a Watch

A brief (and personal) guide

Every watch is unique. Even those made in vast quantities in contemporary factories. Once a watch has been worn, it will pick up traces of its owner's life: the adventures they've been on together, the daily wear, the special occasions it's been brought out for, or even the time it's been sat in its box. This is why, when a watch arrives in our workshop, I start a systematic process of inspections to make sure I capture any and all of the faults it has picked up over the years.

First, I don my watchmaker's loupe, a small magnifying glass that nestles against my eye and enlarges the watch in front of me to three times its size. I carefully check the overall appearance of the watch, looking for marks on the case and signs of water damage on the dial, and whether the winding button (or crown) is worn or shows a telltale flat impact mark, indicating the watch has been dropped on its crown, possibly causing further impact trauma inside. Tiny little scratches on the dial can be a clue that it has been to a less careful watchmaker in the past and might have other repair damage hidden inside. I check that the winding works, that the movement is still running, and that the winding button pulls out and turns the hands freely but not too sloppily; there needs to be some friction, but not too much. By the time I open the case back, I usually have a fairly good idea of what to expect.

The inside of watches' case backs is the prime location to spot the marks of past repairers. Sometimes they're scratched into the metal or scribed in permanent ink. There can be an identifiable name and date, or a code that means nothing to anyone other than the person who left it there. Several marks can indicate that a watch has been regularly sent for servicing over the years, like taking a car for its MOT, and so has been well cared for. Sometimes, however, they forewarn of a string of unresolved issues that it's now my job to unravel.

What happens from here varies from watchmaker to watchmaker, and depends on the calibre of the movement. There is near-infinite variation in the faults I encounter inside a watch, including some surprises I might see just once in my whole career. What follows is how I would approach a very typical manually wound wristwatch from the mid-twentieth century. It can't pretend to be a comprehensive guide, but it's an introduction to the process.

Assuming the watch is running at all, I start by checking it on our timing machine to get a basic indication of performance. Then I remove the bezel, the ring that holds the crystal protecting the dial, allowing me to access it. I pull the winding button into handset and set the time to twelve o'clock, or another time where the hands sit directly over each other. This provides the optimum position to lever the hands up and off without damaging them. To protect the dial, as well as taking great care, I use a thin sheet of plastic while I lift the hands. I repeat the process with the small seconds hand that sits just over six o'clock. With the hands off, I replace the bezel to protect the dial again and turn the watch over.

I slightly loosen the small screw just to the side of the stem of the winding button, allowing me to slide the button and attached stem out of the case. Two screws with large polished heads, slightly domed like the back of a ladybird, hold the movement in its case. I remove these two screws and remove the bezel again to release the movement from the case band, dial first. Putting the case to one side ready for cleaning, I remove the dial by loosening two tiny parallel grub screws that travel horizontally into the sides of the movement, engaging with the feet of the dial hidden inside. Once loose, the dial lifts off, and I place it carefully in a little sealed box along with the hands, to keep them safe while I turn my attention to the mechanism.

The movement is now fully undressed, so I fit it into an adjustable holder, which has sides that can be screwed inwards to hold the edges tightly. Then I lift off the motion work, the wheels that control the rotation of the hands, that has been sandwiched between the movement and dial. I place them to one side in my dust cover. A dust cover is like a horological bento box, with low-walled divisions separating areas in a tray to house the various collections of parts that make up each mechanism within a watch. It is covered by a clear plastic dome with a little handle on the top, a bit like a serving dish, which protects the parts from dust or rolling off my bench.

Now I fit the winding stem, with the button still attached, back into the watch, tightening the little screw to hold it in place again. This makes it easier to handle the movement and allows me to double-check the winding and keyless work now I can see it all moving.

Next I start the mechanical checks. I make sure the fine spiral hairspring is sitting perfectly over the balance wheel, that it's flat and the coils are not bunching to one side. I check the index, which controls the rate of the watch by lengthening or shortening the active length of the hairspring and has settings for 'slow' and 'fast'. It should be aligned in the middle. If the index is cranked all the way over to 'slow' (often marked 'R' for the French 'Retard' meaning 'Delay') it's a clue that the delicate spring might have been replaced or broken and re-pinned in the past and is now too short, making the watch run fast.

I check the whole movement for rust or oil build-up. I may have already spotted both of these faults before I removed the dial. Rust can tint the dial a reddish brown and is usually paired with water staining, while too much oil left by a previous repairer can begin to ooze through the hole in the middle of the dial, leaving a greenish residue and sometimes causing the paint to start lifting. I check for missing parts, which can work loose and end up getting jammed elsewhere in the movement, or are sometimes missing altogether. I switch to a closer magnifying loupe, which enables me to see the tiny components at twenty times their size, and look for cracked ruby jewel bearings that might be wearing the pivots rotating in them. Sometimes, if the watch has been dropped, these pivots break completely, causing the horizontal wheel to flop to one side. I remove any loose chunks of dirt or old jumper fluff that have worked their way into the winding with my tweezers. For general jobs I use my no. 3s, which are as fine at their tip as a sharpened pencil.

Next, I start checking for the faults that are so tiny it's easier to *feel* them than attempt to see them but are nevertheless so significant they could stop a watch altogether. Each wheel inside the watch, from the balance to the train wheels and mainspring barrel, needs to have the right amount of something we call 'shake'. This is the wiggle room that each part needs to function efficiently. Too much wiggle and the depthing (distance between the parts) will be wrong, causing unnecessary wear, variation in timekeeping and wheel teeth skipping through out of control. Too little wiggle and the mechanism will seize up and lock altogether. The right shake is often measured in hundredths, sometimes thousandths, of a millimetre. We check it by holding the part in a pair of finely pointed watchmaker's tweezers and giving it a gentle wobble. It takes a lot of practice but eventually you can feel instinctively whether the shake is right or wrong.

I work through this part of the process in stages. First, I check the shake of the balance. If I'm happy, I remove the balance cock screw and lift it out of the movement, the balance suspended on its hairspring below. I check the balance over, making sure the pivots aren't worn. I check the underside of the wheel for little marks that indicate the staff has been replaced and whether any weight has been filed away to poise it. I remove the jewels and check whether they've started to wear and need to be replaced. By my calculations, a typical watch train running for eighty years would vibrate 12,614,400,000 times. Even something as tough as synthetic ruby can start to erode against the steel staff after this much friction.

Next, I check the depth of the pallets in the teeth of the escape wheel, gently flicking the lever across with the tip of my oiler to make sure the locking is just right. Watchmakers' oilers are like very tiny spatulas made from steel, and are the thickness of a stiff paint-brush bristle. They have a little olive-shaped tip on the end that you dip in oil, the form of the olive holds it in place with surface tension until it's touched against your target. Even without the oil, they are a useful and precise tiny tool and sometimes more gentle than using tweezers. There is a mathematical formula for the perfect angles of depth between the pallets and the escape wheel teeth ('depthing') but, again, tables and charts mean very little when what you're working on is so small. It's easier to go by eye and feel. Depthing is bigger than the shakes, so you can usually see it as well as feel it.*

I'm careful to let any power out of the mainspring before removing the pallets as, once the movement is palletless, there's nothing to stop remaining power whizzing through the train, which in a dirty movement can potentially cause more wear and damage. I pull back the click that engages with the ratchet wheel and stops the mainspring unravelling. Holding the winding button carefully, I slowly let it reverse between my fingers under the releasing power of the spring, allowing the mainspring to unfurl and the power to seep away. I check the shakes of the pallet pivots and remove the bridge securing them. I check the ruby faces of the pallets, to make sure they haven't worn or chipped. I put the bridge, bridge screws and pallets in my dust cover to join the other parts I've placed there so far.

Working through the train, I check it's free and that the shakes are all acceptable. I remove the bridge holding the tops of the

---

* A way of illustrating depthing is if you interlock your right and left hand with straightened fingers, so the middle knuckles align. Give your hands a wiggle, and you'll find you are able to move your fingers, but they still stay firmly locked (correct depthing). However, if you lace your fingers together at their last joint, near the tip, your fingers might slip apart (depthing that is too shallow). If you lace your fingers together at their base you will struggle to wiggle your hands at all (depthing that is too deep).

pivots, which allows me to thoroughly inspect the pivots for wear. I then remove the last and largest bridge to reveal the barrel, removing the ratchet wheel and click to check the shakes. Then, once the barrel is free and in my hand, I check the shakes of the arbor inside, before finally popping the lid off. I remove the arbor and spring and check they are intact and in good condition. In old watches, the mainspring has nearly always fatigued with age and use and needs replacing. Replacing mainsprings can be a balancing act. They need to be the correct height and thickness to give the right power, which is very individual to the watch. Quite often, if you consult the official old charts and use like-for-like measurements the mainspring ends up exerting too much power, as modern springs are made from more efficient steels. Conversely, heavily worn tractors like our old Rebbergs might need all the help with power they can get, requiring a stronger mainspring now than when new. It can involve a bit of trial and error, testing a few springs to make sure you use the right one for the watch rather than the right one by the book. That's something we can only test once the watch is clean and back together, so for now I set it to one side. I remove the winding stem for the last time before cleaning, allowing the wheel and sliding clutch that control the interaction between handset and winding to drop out with it. These are enclosed in a pocket under the barrel bridge so, once exposed, the only thing holding them in the mechanism is the stem they thread onto.

The last set of components to be removed is the keyless work – the parts that allow the watch to be wound without a key and control the action between winding and setting the hands. I flip over the bare movement in its holder, to access the side usually hidden under the dial. The piece that was holding the stem as well as a small wheel, a lever (called a return bar) and a spring all hide under what we call rather literally the 'keyless cover'. I remove the screw and carefully lift this cover, making sure the spring doesn't ping out across the workshop. Return bar springs, often shaped like a shepherd's crook, are a little thicker than the hairspring but still tiny enough to make them a bugger to find! I remove the last parts with my tweezers, storing them away, and set about cleaning out the ruby jewels and bearings by hand with a sharpened piece of wood to make sure any dried old oil is removed. I then put all of the parts I've been carefully storing in my dust cover through our specialist watch cleaner.

Watch cleaners are a bit like dishwashers: you get the best finish if you make sure most of the mess is removed first. So I hand-clean the parts with solvent to remove excess grease and oil before loading them into little steel baskets that clip into the harness of the cleaner. I generally sort the brass parts together in one mini basket and the steel in another (to stop them scratching the brass as they spin in the cleaner) and store any very delicate parts in more individual mini baskets. The harness is like a robotic arm that lowers and lifts the baskets in and out of a series of cleaning solutions or rinses. The last receptacle conceals a heater, which evaporates any remaining rinse out of the movement and leaves the baskets hot to the touch.

While the movement is being cleaned, I turn my attention to the case, cleaning away any dirt and sometimes brightening it with a little polish. If a case is seventy years old or more, it will inevitably look like it's lived a little. I don't try to make cases look 'like new', unless their owner has requested it. I'm sensitive to the fact that those marks can be stories to their owner and they might want them left alone.

Once the parts of the movement are sparkling clean, I reassemble them. I start with the train and barrel, recoiling the mainspring with a special tool and fitting it back in place, laced with a little fresh grease. A slight bit of pressure to the barrel should send the whole train spinning, so it's easy to check that everything is moving freely. If something's locking, there's no point continuing until you've resolved the issue.

Now I reassemble the keyless work, carefully loading the little spring in place before capturing it under the cover before it pings away. It's a bit like trying to catch a grasshopper under cupped hands in the garden. I apply a little grease or oil to all the jewelled bearing surfaces and points of friction, careful to use enough to do the job but not so much it oozes onto the dial or anywhere else it shouldn't be. I check the keyless work is all functional, clicking between handset and winding. When I try winding the watch the train should now whizz through freely.

I lock the train with the pallets, retuning them to their position engaging with the escape wheel teeth. I wind the watch, turning the button once or twice, and watch the lever suddenly jump into position, held by the pressure now backed up in the train. I check the lever is flicking across in a lively fashion before moving on to the final stage.

I chance winding the mainspring right the way up to full power and return the assembled balance with its freshly oiled jewels to the movement. This is the moment when the watch starts to tick again. It's particularly poignant after the long restoration of a watch that has been silent for many years. It's when the watch comes back to life, its little escapement starting to beat and its hairspring methodically breathing in and out again.

I can tell by sight how healthy a balance is by the degree of its oscillations. In a watch of this age and style, I like it to be between 280 degrees and 300 degrees. Any greater and the balance risks swinging right the way round and knocking the wrong side of the lever, creating a tick that sounds a bit like a galloping horse. This can happen if your new mainspring is too strong. If the angle is too small, the watch is susceptible to positional errors, gaining or losing with the movement of the wearer. I wait until I can tell by eye that it's in the right ballpark before putting the watch anywhere near our timing machine, which reads the ticks to tell us exactly how the watch is performing.

The timing machine allows us to check the poising – that the weight of the balance wheel is evenly distributed. If one spot is too heavy, the reading on the machine will show a loss when the balance is on its side (the position it would be in if you were wearing it with your arm by your side), and a gain when the movement is rotated the other way. This impacts timekeeping. If the balance is out of poise, I remove it and roll the wheel on a set of sharp ruby jaws that allow it to roll freely until the heavy spot drops to the bottom. Once the culprit is identified, I remove a tiny scraping of metal with a cutter (these are the marks you can see if someone has done this before) to reduce the weight, and try rolling it again. And again. And again. Until it rolls free and comes to a slow stop without swinging. Sometimes I get this right first time, and sometimes it takes hours of patience to achieve.

When the power is off, the ruby pin (known as an impulse jewel) under the balance that flicks the pallet lever back and forth should rest centrally in the fork of the lever; we call this being in beat. With the other checks completed, the movement is ready for casing up. I replace the motion work and dial, securing those little grub screws to hold the dial securely. I replace the hands, checking that they align perfectly at the hour. I return the movement to the case band, removing the stem one last time and threading it back through the hole in the case before securing it tightly. I replace the last two screws holding the movement in the band and replace the back.

The last test is a practical one, or as close to practical as we can achieve in the workshop. We put the watch onto the 'wrist' of a machine that rotates it through all positions as it runs, checking it continues to perform while

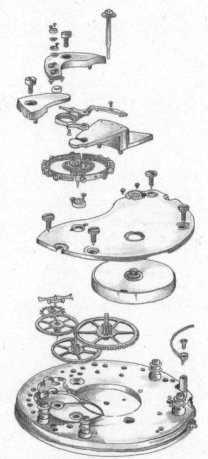

An exploded pocket-watch movement.

being moved around like it will be in wear. We test it in movement during the day, and rest it in one position at night, over the course of seven days, ticking away non-stop. Only when we're satisfied that the watch is behaving itself do we reunite it with its owner.

Glossary

Arbor is a rod or parallel spindle onto which a part that needs to rotate or pivot is mounted, for example a watch train wheel.

Automatic, also called 'self-winding'. A watch with a mechanism that winds the mainspring using the movement of the wearer, usually using the swinging or rotating motion of a weight.

Atomic time is a high-precision form of timekeeping generated by atomic clocks. An increase in energy can cause electrons orbiting the atomic nucleus to jump to a higher orbit. As the electrons jump, they emit energy at a specific oscillation frequency that can be used for timekeeping.

Balance spring (see hairspring)

Balance bridge is the name of the component fitted to the top plate of a watch with a verge escapement (occasionally early lever and cylinder escapements), designed to hold the top pivot of the balance staff. It consists of a round plate, or table, which is often decorated with piercing and/or engraving and attached to the top plate by two feet secured by two screws forming the shape of a bridge.

A pierced and engraved balance cock with Green Man design, found in an English watch from the 1760s

Balance cock is as above, only held by a single foot and screw.

Balance staff is the central component, or backbone, of the balance assembly. The balance wheel, impulse roller and collet in the middle of the hairspring are all friction-fitted to this central staff, which oscillates on pivots at the top and bottom. Variations of the balance staff form the basis of every single watch balance assembly invented.

Balance is an oscillating wheel in the movement of a watch, similar in appearance to the design of a steering wheel, and is responsible for regulating the release of power from the mainspring. It is friction-mounted and then riveted to the balance staff.

Bezel is the front part of a watch case in which the transparent glass (or mineral crystal or sapphire) is secured to reveal the dial underneath.

Blued steel is steel that has been heat treated, traditionally using the flame of a spirit lamp, to create an oxide layer that makes it appear blue in colour. As the steel is heated it changes colour, first turning a straw-yellow shade before going deep purple, then midnight blue. This blue gradually lightens in colour to a bright, almost electric blue before fading back to a grey steel colour. The oxide layer offers some protection from rust and tempers hard steel, but in watch-making it is also performed purely for aesthetic reasons. Watchmakers choose their preferred shade and stop the colour transition by removing the steel from the heat.

A glass watchmaker's spirit lamp, used for jobs like heat-bluing steel.

Champlevé is a style of solid silver or gold dial, often with engraved numerals in-filled with black enamel or wax, and typically decorated with ornate engraving, piercing and chasing. The modern term also refers to dials that are engraved, then covered with translucent vitreous enamel.

Chronometer is the term given to a watch or clock capable of operating to the very highest standards of accuracy. The title is awarded by national independent governing bodies who test them. Historically, this occurred in observatories such as Kew (UK), Besançon (France), and Neuchâtel and Geneva (Switzerland). The current main assessment centre is the Contrôle Officiel Suisse des Chronomètres (or COSC). Accuracy is measured over many days (fifteen at COSC; Kew used to have a forty-four-day trial) and through a range of different temperatures and positions. Only watches that meet these exacting standards are allowed to be referred to as chronometers.

Chronograph is a timepiece that also has a stopwatch function that can start, stop and reset without interfering with the timekeeping element of the watch.

Complications are functions that go beyond any device used to improve timekeeping, such as a chronograph, calendar display or repeating work.

Crown, also known as the button, is the part on the side of the case that can be turned to wind up or set the time on a watch.

Depthing is the positional relationship between two interacting components. Optimal depthing ensures the parts work efficiently together without slipping past each other (which means the depthing is too shallow) or locking or generating too much friction (which means the depthing is too deep).

Ébauche is the term given to a standardised movement supplied to order in a complete but unfinished state with the intention that the purchaser can customise, finish and sign it accordingly before adding their own case, dial and hands. *Ébauche* manufacture was perfected in the mid-nineteenth century.

Engine turning is a style of engraving performed on a rose or straight-line engine machine. The finished result is geometric in form, resembling a design a spirograph might create, or a weave like the side of a wicker basket.

English lever is a type of escapement invented in England and used there from the second half of the eighteenth century to the beginning of the twentieth century. It differs from the Swiss lever in the shape of the teeth of the escape wheel, which are long, fine and pointed.

Equation of time is the difference between mean solar time and apparent solar time, meaning the difference between the time shown on a watch or clock and the time shown on a sun or ring dial.

Escapement is the collective name given to the group of components in a watch responsible for controlling the release of power from the mainspring and reducing the speed the train wheels rotate to one useable for timekeeping.

Établissage is the early production-line manufacturing of non-standardised blank watch movements (evolved into the *ébauche*), which was developed in Switzerland from the start of the eighteenth century.

Form watch is the term given to a late-sixteenth or early-seventeenth-century watch that is, quite literally, created in the form of something else. Popular forms included flowers, animals, skulls and religious iconography.

Fusee is a device used to achieve uniform torque from the mainspring. When the mainspring is at full wind it exerts a stronger force than when the watch is nearly unwound. The force is evened out using a gut line (or later chain), which transmits the power to a graduated barrel known as the fusee. When the mainspring is at full wind, the line turns the smallest diameter of the fusee barrel, reducing its power. The line works its way down the graduation, inverting the mainspring's power against the fusee diameter.

Gilding is the predecessor to gold plating, also referred to as 'fire' or 'wash' gilding. An amalgam of gold and other metals and chemicals in the form of a paste would be painted onto the surface of silver, or a base metal like copper or brass. The liquid was burned off with a flame, leaving the gold baked on to the surface. The fumes created were incredibly dangerous as the process uses ingredients such as mercury, ammonia and nitric acid. Gilding was replaced by electroplating.

Greenwich meridian is the imaginary line that runs through Greenwich in London, UK, opposite to the international date line.

Hairspring, also known as a balance spring, is the fine spring formed in a spiral that regulates the speed at which the balance oscillates. When shortened the watch runs faster and when lengthened it runs more slowly. It is secured to the balance staff by a collet at its centre. In early watches it was made of a hog's-hair bristle.

Jewels are synthetic corundum (ruby or sapphire) pivot bearings. Corundum is used for its exceptionally hard-wearing properties. These replaced brass bushes and start to appear in the eighteenth century.

Keyless work is the group of components that work together to allow a watch to be wound and control the transition between winding and setting the hands (by pulling out the crown, on modern watches). During handset, the keyless work engages with the motion work.

Latitude is a position calculated using the series of imaginary lines on a map or globe that run from east to west, parallel to the equator.

Longitude is a position calculated using the series of imaginary vertical lines that run at 15-degree intervals from the North to South Pole on a globe and are counted from the Greenwich meridian.

Loupe is a magnifying lens that, in watchmaking, is worn against the eye. For general work a three-times magnification is used, although for more detailed inspections a range of loupes are used, generally up to around twenty-times magnification.

A loupe.

Mainspring is the power source in a mechanical watch. It is the name given to the coiled ribbon-like spring contained within a barrel (the mainspring barrel) that can be wound up. The wound spring creates a rotating force as it releases, which is transferred down a series of toothed wheels to the escapement, where the speed of release is regulated.

Masstige is an amalgamation of the terms 'mass produced' and 'prestige', used to refer to the commercialisation of luxury objects.

Motion work is the series of wheels and pinions that control the movement of the watch hands. The hour wheel is geared to turn once every twelve hours, and the canon pinion once every sixty minutes. In normal operation, their movement is controlled by the train wheels. In handset, the keyless work overrides this to allow manual handsetting.

Oscillation is the regular back-and-forth movement, or rhythm, of an object such as a pendulum or wheel.

Pallets (entrance and exit) are the parts of the escapement that allow the catch and release of one tooth at a time of the escape wheel. This regular staggered catch/release controls the release of power from the mainspring at a rate that can be used for timekeeping.

Pare-chute, **or parachute,** is the name of the first type of shock absorber introduced in a watch mechanism in 1790. It was invented in Paris by Abraham-Louis Breguet and helped protect the delicate pivots of the balance staff from breaking in the event that the watch was knocked or dropped.

Pair-cases is the name given to the design of watch in which the movement is housed in an inner case, which is then protected within a further outer case.

Piezoelectricity, from the Greek for pressure or push, in this context refers to the small electrical pulse produced by a quartz crystal when it is subjected to mechanical stress.

Pivots are the narrow terminal points of an arbor, which are smaller in diameter, and highly polished to reduce friction as much as possible. They are the points the arbor rotates or pivots on.

Positional error is the variation in timekeeping caused by changes in the direction of gravitational pull as the watch moves through a range of positions, as it does during wear.

Quartz crystal regulates the electronic oscillator in a quartz-battery-powered wristwatch using its piezoelectric properties.

Raising is a process in which metal is repeatedly hammered and then annealed (softened by heating) to stretch and form it into a shape such as a bowl, cup or watch case.

Repeater is a mechanism in a timepiece that chimes hours and quarter-hours, or hours, minutes and quarter-hours, on gongs made from wire or small bells.

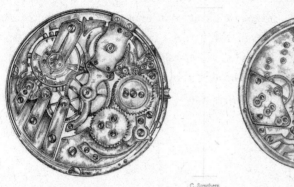

Louis Audemars Quarter Repeater

C. Struthers
Birmingham, B1.

A quarter-repeating pocket watch movement made in Switzerland by Louis-Benjamin Audemars in around 1860.

Repoussé is a technique used in silversmithing where a design is punched into the back of a piece of metal to create a relief, before being engraved and chased from the front to refine the detail. Commonly used on the outer case of pair-cased watches in the eighteenth century.

Shake is the amount of end and sideways movement a pivot has in its bearing. Too much movement will interfere with depthing and too little will cause the part to seize.

Shock setting is the general term for the components that help to protect pivots from breaking in the event of an impact. This includes the *pare-chute*, the concept of which has been refined and redesigned over the centuries to include a wide range of systems.

Staff (see balance staff)

Temperature compensation refers to the range of different inventions developed to compensate for the variations caused by thermal expansion and contraction.

Tourbillon is a device invented by Abraham-Louis Breguet in 1801 that reduces positional error by keeping the escapement and balance assembly constantly rotating through 360 degrees.

Train wheels are the series of toothed wheels and pinions that interconnect in a watch movement to gear down the power from the mainspring to rotation speeds compatible with timekeeping.

Verge is the name of the earliest escapement found in watches. First used in clocks and phased out at the turn of the nineteenth century, the mechanism consists of a balance staff with two flags positioned at right angles. The staff, secured to the oscillating balance, rotates back and forth, allowing the pallets to release one tooth at a time of an engaging wheel, referred to as the escape or crown wheel.

Acknowledgements

This book brings together so much of my life, career and education that I could write a list of equal length for every person who has helped me on my way. But I'll try to keep it short.

For introducing me to the world of watchmaking I'd like to thank my tutors Paul Thurlby and Jim Kynes. As a jeweller I'm indebted to the tuition of Peter Ślusarczyk and Eimear Conyard. As a historian, my studies wouldn't have been possible without the support of Dr Lawrence Green and Professor Kenneth Quickenden.

This book came about at a moment when, after many submissions and rejections, I'd given up all hope of publishing it. Thanks to Kirsty McLachlan for finding me and encouraging me to give it another go, to my agent David Godwin, and to my editor Kirty Topiwala for commissioning this book, along with her wonderful team at Hodder: Rebecca Mundy, Jacqui Lewis, Tom Atkins, Helen Flood and my picture editor Jane Smith. Holly Ovenden designed the beautiful cover. Roma Agrawal has been my informal mentor, keeping me on the straight and narrow. I must also take a moment to honour the long-suffering Victoria Millar who helped me to weave together the multitude of strands of this story. Thank you for your late nights and cheer-up cat photos when things have been a struggle.

For assisting with my research enquiries and helping me check the details of this vast subject, thanks to (in no particular order) Justin Koullapis, Alom Shaha, Dr Michelle Bastian, Professor Kevin

Birth, Michael Clerizo, Mike Cardew, Dr Richard Hoptroff, Dr Stephanie Davies, David Goodchild, Dr Jim Beveridge, Karen Bennett, Chantal Bristow, Professor Francesco D'Errico, Dr Katie Russell-Friel, Mr Ronald Mifsud, Anna Rolls, Elizabeth Doerr, Seth Kennedy, David Barrie, Mollie Hughes, Mike Frayn and Dr James Fox. Thanks also to Professor Joe Smith and Professor Renata Tyszczuk from Smith of Derby for showing me around their epic workshop.

Thank you to the ultra-talented Andy Pilsbury who photographed all of the images in this book as well as the majority of our photographs over the years. We're lucky to have such skilled friends. Andy welcomed his first child, Poppy Pilsbury, in 2022 so I'm going to take this moment to immortalise her in print and say welcome to the world, Poppy! I'm also grateful to Jen O'Shaugnessy for her photo-editing skills, and to the owners and caretakers of the watches and objects included whose support of the book made these images happen.

Much of my research would not have been possible without the support of museums and their curators. I particularly want to thank David Thompson, Paul Buck, Oliver Cooke and Laura Turner of the British Museum; Alan Midleton, Alex Bond, Izzy Davidson, Dr Robert Finnegan and Dave Ellis from the Museum of Timekeeping; Anna Rolls from the Worshipful Company of Clockmakers; Dr David Morris from the McGregor Museum; and Amy Taylor from the Ashmolean Museum. The thorough list of museums with horological objects and collections all over the world was greatly enriched by ideas from my amazing followers on Instagram; thank you all for your kind suggestions.

There have been a number of educational bodies and charities that have supported us and our workshop over the years. I mention them here both to thank them, and in case anyone reading this would also like to find out more about the traditional crafts described in this book or learn about the art of watchmaking: the British Horological Institute, Queen Elizabeth Scholarship Trust, Heritage Crafts and the Association of Heritage Engineers.

I'd like to thank the creative cluster of fellow makers with whom I've had the pleasure of collaborating, and who make what we do possible: Henry Deakin, The Wizard (a.k.a. Steve Crump), Dave Fellows, Andrew Black, Anita Taylor, Liam Cole, Sally Morrison, Lewis Heath, Florian Güllert, Mike Couser, Neil Vasey, Anousca Hume, Gabi Gucci, May Moorhead, Callum Robinson and Marisa Giannasi. And special thanks to our exceedingly patient friend and director Jan Lawson for having the astute business mind that Craig and I lack. I doubt we'd still have a workshop to write about without you.

And for keeping me well and able to continue watchmaking, my thanks to Sharon Letissier and Dr Niraj Mistry.

To my soulmate, inspiration and illustrator Craig Struthers, thank you for your unwavering support while I've been writing this book, and over the last two decades. And my four-legged family of waifs and strays, my short and furries, for keeping me on my toes and being there for a cuddle whenever I've needed one: watch dog Archie, cats Alabama and Isla, and our mouse Morrissey. And, of course, to my parents and family, I am who I am today because of you, so when I'm a pain in the arse you have no one to blame but yourselves.

This book is dedicated to the memory of two great losses that occurred during its writing. Adam Phillips was the last independent watch casemaker in the UK who took Craig on for an informal apprenticeship in 2017 to share his skills. His generosity, knowledge and kindness will not be forgotten. Neither will Craig forget the excuse Adam provided him with for always needing new (old) tools. One can 'never be over-lathed'. As a fellow cat-lover, I hope Adam would have been happy to share his dedication with my loyal old friend Indy, who sat on my lap for most of the time I was writing this book. She passed away the day after I submitted my manuscript. She was a very fine cat, a very fine cat indeed.

Picture credits

p. 1 (top): courtesy of the McGregor Museum, Kimberley, South Africa. Bone photograph © F. D'Errico and L. Backwell/McGregor Museum/Andy Pilsbury. Digital editor: Jen O'Shaughnessy.

p. 1 (bottom left and right): courtesy of the Clockmakers' Museum (at the Science Museum, London, UK).

p. 2 (top and bottom): courtesy of the Ashmolean Museum, Oxford, UK.

p. 5 (top and bottom): courtesy of the Clockmakers' Museum (at the Science Museum, London, UK).

p. 6: courtesy of the Museum of Timekeeping, Newark, UK.

p. 7 (top and bottom): by permission of James Dowling.

p. 8: by permission of Kevin Carter (@kccarter1952).

Bibliography

Abulafia, D. (2019). *The Boundless Sea: A Human History of the Oceans*. Allen Lane, London.

Albert, H. (2020). 'Zoned out on timezones.' *Maize*, 30 January. Available at: https://www.maize.io/magazine/timezones-extreme-jet-laggers (Accessed 12 May 2021).

Álvarez, V.P. (2015). 'The Role of the Mechanical Clock in Medieval Science'. *Endeavour*, 39 (1), pp. 63–8.

Anon (1772). *A View of Real Grievances, with Remedies Proposed for Redressing Them*. London.

Anon (1898). *The Reign of Terror, a Collection of Authentic Narratives of the Horrors Committed By the Revolutionary Government of France Under Marat and Robespierre*. J.B. Lippincott Company, Philadephia.

Antiquorum (1991). *The Art of Breguet*. Habsburg Fine Art Auctioneers. Sale catalogue 14 April. Schudeldruck, Geneva.

Baker, A. (2012). '"Precision", "Perfection", and the Reality of British Scientific Instruments on the Move during the 18th Century'. *Material Culture Review*, 74–5 (Spring), pp. 14–28.

Baker, S.M. and Kennedy, P.F. (1994). 'Death by Nostalgia: A Diagnosis of Context-Specific Cases'. *NA – Advances in Consumer Research*, vol. 21, eds. Chris T. Allen and Deborah Roedder John, Provo, UT: Association for Consumer Research, pp. 169–74.

Balmer, R.T. (1978). 'The Operation of Sand Clocks and Their Medieval Development'. *Technology and Culture*, 19 (4), pp. 615–32.

Barrell, J. (1980). *The Dark Side of the Landscape: The Rural Poor in English Painting*. Cambridge University Press, Cambridge.

Barrie, D. (2015). *Sextant: A Voyage Guided by the Stars and the Men Who Mapped the World's Oceans*. William Collins, London.

Barrie, D. (2019). *Incredible Journeys: Exploring the Wonders of Animal Navigation*. Hodder & Stoughton, London.

Bartky, I. (1989). 'The Adoption of Standard Time'. *Technology and Culture*, 30 (1), pp. 25–56.

Baxter, R. (1673). *A Christian directory, or, A summ of practical theologie and cases of conscience directing Christians how to use their knowledge and faith, how to improve all helps and means, and to perform all duties, how to overcome temptations, and to escape or mortifie every sin: in four parts . . . / by Richard Baxter*. Printed by Robert White for Nevill Simmons, London.

Beck, J. (2013). 'When Nostalgia Was a Disease'. *The Atlantic*, 14 August. Available at: https://www.theatlantic.com/health/archive/2013/08/when-nostalgia-was-a-disease/278648 (Accessed 14 May 2021).

Betts, J. (2020). *Harrison*. National Maritime Museum, London.

Birth, K. (2014). 'Breguet's Decimal Clock'. The Frick Collection, *Members' Magazine*, Winter.

Breguet (2021). *'Grande Complication' pocket watch number*. Available at: https://www.breguet.com/en/house-breguet/manufacture/marie-antoinette-pocket-watch (Accessed 18 May 2021).

Breguet (2021). *1810, The First Wristwatch*. Available at: https://www.breguet.com/en/history/inventions/first-wristwatch (Accessed 18 May 2021).

Breguet, C. (1962). *Horologer*. Translated by W.A.H. Brown. E.L. Lee, Middlesex.

Centre, J.I. (2021). 'Bacteria Can Tell the Time with Internal Biological Clocks'. *Science Daily*, 8 January. Available at: https://scitechdaily.com/bacteria-can-tell-the-time-with-internal-biological-clocks (Accessed 22 April 2021).

Chapuis, A. and Jaquet, E. (1956). *The History of the Self-Winding Watch 1770–1931*. B.T. Batsford Ltd, London.

Chapuis, A. and Jaquet, E. (1970). *Technique and History of the Swiss Watch*. Translated ed. Hamlyn Publishing Group Limited, Middlesex.

Chevalier, J. and Gheerbrant, A. (1996). *Dictionary of Symbols*. Translated 2nd ed. Penguin, London.

Church, R.A. (1975). 'Nineteenth-Century Clock Technology in Britain, the United States, and Switzerland'. *Economic History Review*, New Series, 28[4].

Clarke, A. (1995). *The Struggle of the Breeches: Gender and the Making of the British Working Class*. University of California Press, Berkley.

Clarke, A. (2020). 'Edinburgh's iconic Balmoral Hotel clock will not change time at New Year'. *Edinburgh Live*, 29 December. Available at: https://www.edinburghlive.co.uk/news/edinburgh-news/edinburghs-iconic-balmoral-hotel-clock-19532113?utm_source=facebook.com&utm_medium=social&utm_campaign=sharebar&fbclid=IwAR0HxWdnV5H4VrQT51OofOkUMWs_kXaHMo_h4LvHCu2Fr1PFsLTgfl6Qono (Accessed 5 May 2021).

Clayton (1755). *Friendly Advice to the Poor; written and published at the request of the late and present Officers of the Town of Manchester*.

Corder, J. (2019). 'A look at the new $36,000 1969 Seiko Astron'. *Esquire*, 6 November. Available at: https://www.esquireme.com/content/40676-a-look-at-the-new-36000-1969-seiko-astron-draft (Accessed 14 May 2021).

Cummings, G. (2010). *How the Watch Was Worn: A Fashion for 500 Years*. The Antique Collectors' Club, Suffolk.

Cummings, N. and Gráda, C.Ó. (2019). 'Artisanal Skills, Watchmaking, and the Industrial Revolution: Prescot and Beyond'. Competitive Advantage in the Global Economy (CAGE) Online Working Paper Series 440. Available at: https://ideas.repec.org/p/cge/wacage/440.html (Accessed 8 April 2021).

Daniels, G. (1975). *The Art of Breguet*. Sotheby's Publications, London.

Darling, D. (2004). *The Universal Book of Mathematics: From Algebra to Zeno's Paradoxes*. John Wiley & Sons, New Jersey.

Davidson, H. (2021). 'Tiananmen Square watch withdrawn from sale by auction house'. *Guardian*, 1 April. Available at: https://

www.theguardian.com/world/2021/apr/01/tiananmen-square-watch-given-chinese-troops-withdrawn-from-sale-fellows-auction-house (Accessed 14 May 2021).

Davie, L. (2020). 'Border Cave finds confirm cultural practices'. *The Heritage Portal*. Available at: http://www.theheritageportal.co.za/article/border-cave-finds-confirm-cultural-practices (Accessed 6 July 2020).

Davis, A.C. (2016). 'Swiss Watches, Tariffs and Smuggling with Dogs'. *Antiquarian Horology*, 37 (3), pp. 377–83.

D'Errico, F.; Backwell, L.; Villaa, P.; Deganog, I.; Lucejkog, J.J.; Bamford, M.K.; Highamh, T.F.G.; Colombinig, M.P. and Beaumonti, P.B. (2012). 'Early Evidence of San Material Culture Represented by Organic Artifacts from Border Cave, South Africa'. *Proceedings of the National Academy of Sciences of the United States of America*, 14 August, 109 (33), pp. 13, 214–13, 219.

D'Errico, F., Doyon, L., Colagé, I., Queffelec, A., Le Vraux, E., Giacobini, G., Vandermeersch, B., Maureille, B. (2017). 'From Number Sense to Number Symbols. An Archaeological Perspective'. *Philosophical Transactions of the Royal Society*. B 373: 20160518.

De Solla Price, D. (1974). 'Gears from the Greeks: The Antikythera Mechanism – A Calendar Computer from ca. 80 B.C.' *Transactions of the American Philosophical Society*, 64 Pt. 6. Philadelphia.

Dickinson, H.W. (1937). *Matthew Boulton*. Cambridge University Press, Cambridge.

Diop, C.A. (1974). *The African Origin of Civilization: Myth or Reality*. Chicago Review Press, Chicago.

Dohrn-van Rossum, G. (1996). *History of the Hour: Clocks and Modern Temporal Orders*. Translated ed. The University of Chicago Press, Chicago.

Dowling, J. and Hess, J.P. (2013). *The Best of Time: Rolex Wristwatches: An Unauthorised History*. 3rd ed. Schiffer Publishing Ltd, Pennsylvania.

Dyke, H. (2020). *Our Experience of Time in the Time of Coronavirus*

Lockdown, Cambridge Blog. Available at: http://www.cambridge-blog.org/2020/05/our-experience-of-time-in-the-time-of-coronavirus-lockdown (Accessed 11 February 2021).

Erickson, A.L. (Unpublished). *Clockmakers, Milliners and Mistresses: Women Trading in the City of London Companies 1700–1750*. Available at: https://www.campop.geog.cam.ac.uk/research/occupations/outputs/preliminary/paper16.pdf

Evers, L. (2013). *It's About Time: From Calendars and Clocks to Moon Cycles and Light Years – A History*. Michael O'Mara Books Ltd, London.

Falk, D. (2008). *In Search of Time: The Science of a Curious Dimension*. St. Martin's Press, New York.

Forster, J. and Sigmond, A. (2020). *Accutron: From the Space Age to the Digital Age*. Assouline Collaboration.

Forsyth, H. (2013). *London's Lost Jewels: The Cheapside Hoard*. Philip Wilson Publishers Ltd, London

Forty, A. (1986). *Objects of Desire: Design and Society since 1750*. Cameron Books, Dumfriesshire.

Foulkes, N. (2019). 'The Independent Artisans Changing the Face of Watchmaking'. *Financial Times*, How to Spend It, 12 October.

Foulkes, N. (2019). *Time Tamed: The Remarkable Story of Humanity's Quest to Measure Time*. Simon & Schuster, London.

Fraser, A. (2018). *Mary, Queen of Scots*. Fiftieth-anniversary ed. Weidenfeld & Nicolson, London.

Freeman, S. (2021). 'Parents find time passes more quickly, researchers reveal.' *The Times*, 22 February. Available at: https://www.thetimes.co.uk/article/parents-find-time-passes-more-quickly-researchers-reveal-sqvvod65v (Accessed 22 June 2022)

Fullwood, S. and Allnutt, G. (2017–present). The AHS *Women and Horology* Project. Available at: https://www.ahsoc.org/resources/women-and-horology/ (Accessed 18 May 2021).

Ganev, R. (2009). *Songs of Protest, Songs of Love: Popular Ballads in Eighteenth Century Britain*. Manchester University Press, Manchester.

Geffen, Anthony (director) (2010). *The Wildest Dream* (film). United States, Altitude Films with Atlantic Productions.

Glasmeier, A.K. (2000). *Manufacturing Time: Global Competition in the Watch Industry, 1795–2000*. The Guilford Press, London.

Glennie, P. & Thrift, N. (2009). *Shaping the Day: A History of Timekeeping in England and Wales 1300–1800*. Oxford University Press, Oxford.

Good, R. (1965). 'The Mudge Marine Timekeeper'. *Pioneers of Precision Timekeeping: A Symposium*. Antiquatian Horological Society, London.

Gould, J.L. (2008). 'Animal Navigation: The Longitude Problem'. *Current Biology*, 18 (5), pp. 214–216.

Guye, S. and Michel, H. (1971). *Time & Space: Measuring Instruments from the 15th to the 19th Century*. Pall Mall Press, London.

Gwynne, R. (1998). *The Huguenots of London*. The Alpha Press, Brighton.

Hadanny, A.; Daniel-Kotovsky, M.; Suzin, G.; Boussi-Gross, R.; Catalogna, M.; Dagan, K.; Hachmo, Y.; Abu Hamed, R.; Sasson, E.; Fishlev, G.; Lang, E.; Polak, N.; Doenyas, K. et al. (2020). 'Cognitive Enhancement of Healthy Older Adults Using Hyperbaric Oxygen: A Randomized Controlled Trial'. *Aging* (Albany, NY), 12 (13), pp. 13740–13761.

Häfker, N. S.; Meyer, B.; Last, K.S.; Pond, D.W.; Hüppe, L.; Teschke, M. (2017). 'Circadian Clock Involvement in Zooplankton Diel Vertical Migration'. *Current Biology*, 27 (14), (24 July), pp. 2194–2201.

Heaton, H. (1920). *The Yorkshire Woollen and Worsted Industries, from the Earliest Times up to the Industrial Revolution*. Clarendon Press, Oxford.

Helfrich-Förster, C., Monecke, S., Spiousas, I., Hovestadt, T., Mitesser, O. and Wehr, T.A. (2021). 'Women Temporarily Synchronize Their Menstrual Cycles with the Luminance and Gravimetric Cycles of the Moon'. *Science Advances*, 7, eabe1358.

Hom, A. (2020). *International Relations and the Problem of Time*. Oxford University Press, Oxford.

House of Commons (1817). *Report from the Committee on the Petitions of Watchmakers of Coventry*. London, 11 July.

House of Commons (1818). *Report from the Select Committee Appointed to Consider the Laws Relating to Watchmakers.* London, 18 March.

James, G.M. (2017). *Stolen Legacy: The Egyptian Origins of Western Philosophy.* Reprint ed., Allegro Editions.

Jones, A.R. and Stallybrass, P. (2000). *Renaissance Clothing and the Materials of Memory.* Cambridge University Press, Cambridge.

Jones, M. (1990). *Fake? The Art of Deception.* British Museum Publications, London.

Jones, P.M. (2008). *Industrial Enlightenment: Science, Technology and Culture in Birmingham and the West Midlands 1760–1820.* Manchester University Press, Manchester.

Keats, A.V. (1993). 'Chess in Jewish History and Hebrew Literature'. University College, University of London, PhD thesis.

Klein, M. (2016). 'How to Set Your Apple Watch a Few Minutes Fast'. *How-To Geek.* Available at: https://www.howtogeek.com/237944/how-to-set-your-apple-watch-so-it-displays-the-time-ahead (Accessed 8 February 2021).

Landes, D. (1983). *Revolution in Time: Clocks and the Making of the Modern World.* Harvard University Press, Massachusetts.

Lardner, D. (1855). *The Museum of Science and Art, Vol. 6,* Walton & Maberly, London.

Lester, K. and Oerke, B.V. (2004). *Accessories of Dress: An Illustrated Encyclopaedia.* Dover Publications, New York.

Locklyer, J.N. (2006). *The Dawn of Astronomy: A Study of Temple Worship and Mythology of the Ancient Egyptians.* Dover Edition. Dover Publications, New York.

Lum, T. (2017). 'Building Time Through Temporal Illusions of Perception and Action: Sensory & Motor Lag Adaption and Temporal Order Reversals'. Vassar College, thesis, p. 6. Available at: https://s3.us-east-2.amazonaws.com/tomlum/Building+Time+Through+Temporal+Illusions+of+Perception+and+Action.pdf (Accessed 19 April 2021).

Marshack, A. (1971). *The Roots of Civilization.* McGraw-Hill, New York.

Masood, E. (2009). *Science & Islam: A History.* Icon Books Ltd, London.

Mathius, P. (1957). 'The Social Structure in the Eighteenth Century: A Calculation by Joseph Massie'. *Economic History Review* (Second Series), X (1) pp. 30–45.

Matthes, D. (2015). 'A Watch by Peter Henlein in London?' *Antiquarian Horology*, 36 [2] (June 2012), pp. 183–94.

Matthes, D. and Sánchez-Barrios, R. (2017). 'Mechanical Clocks and the Advent of Scientific Astronomy'. *Antiquarian Horology*, 38 (3), pp. 328–42.

May, W.E. (1973). *A History of Marine Navigation*. G.T. Foulis, London.

Mills, C. (2020). 'The Chronopolitics of Racial Time'. *Time & Society*, 29 (2), pp. 297–317.

Moore, K. (2016). *The Radium Girls*. Simon & Schuster, London.

Morus, I.W. (ed.) (2017). *The Oxford Illustrated History of Science*. Oxford University Press, Oxford.

Mudge, T. (1799). *A Description with Plates of the Time-keeper Invented by the Late Mr. Thomas Mudge*. London.

Murdoch, T.V. (1985). *The Quiet Conquest: The Huguenots, 1685 to 1985*. Museum of London, London.

Murdoch, T.V. (2022). *Europe Divided: Huguenot Refugee Art and Culture*. V&A, London.

Myles, J. (1850). *Chapters in the Life of a Dundee Factory Boy, an Autobiography*. Adam & Charles Black, Edinburgh.

Neal, J.A. (1999). *Joseph and Thomas Windmills: Clock and Watch Makers; 1671–1737*. St Edmundsbury Press, Suffolk.

Newberry, P.E. (1928). 'The Pig and the Cult-Animal of Set'. *The Journal of Egyptian Archaeology*, 14 (3/4), 211–225.

Newman, S. (2010). *The Christchurch Fusee Chain Gang*. Amberley Publishing, Stroud.

Oestmann, G. (2020). 'Designing a Model of the Cosmos'. In *Material Histories of Time: Objects and Practices, 14th–19th Centuries*. Bernasconi, G. and Thürigen, S. (eds.). Walter de Gruyter, Berlin, pp. 41–54.

Payne, E. (2021). 'Morbid Curiosity? Painting the Tribunale della Vicaria in Seicento Naples' (lecture, Courtauld Research Forum, 3 February 2021.)

Peek, S. (2016). 'Knocker Uppers: Waking up the Workers in Industrial Britain'. BBC, 27 March. Available at: https://www.bbc.co.uk/news/uk-england-35840393 (Accessed 10 January 2021).

Popova, M. (2014). 'Why Time Slows Down When We're Afraid, Speeds Up as We Age, and Gets Warped on Vacation'. *The Marginalian*. 15 July. Available at: https://www.themarginalian. org/2013/07/15/time-warped-claudia-hammond (Accessed 16 September 2022)

Quickenden, K. and Kover, A.J. (2007). 'Did Boulton Sell Silver Plate to the Middle Class? A Quantitative Study of Luxury Marketing in Late Eighteenth-Century Britain.' *Journal of Macromarketing*, 27 (1), pp. 51–64.

Rameka, L. (2016). 'Kia whakatōmuri te haere whakamua: I walk Backwards into the Future with My Eyes Fixed on My Past.' *Contemporary Issues in Early Childhood*, 17 (4), pp. 387–98.

Ramirez, A. (2020). *The Alchemy of Us: How Humans and Matter Transformed One Another*. The MIT Press, Cambridge, Massachusetts.

Rees, A. (ed.) (1820). *The Cyclopaedia, or Universal Dictionary, Vol. 2*. Longman, Hurst, Rees, Orme, and Brown, London.

Ribero, A. (2003). *Dress and Morality*. B.T. Batsford, London.

Roe, J.W. (1916). *English and American Tool Builders: Henry Maudslay*. McGraw-Hill, New York.

Rolex (2011). *Perpetual Spirit: Special Issue – Exploration*. Rolex SA, Geneva.

Rooney, D. (2008). *Ruth Belville: The Greenwich Time Lady*. National Maritime Museum, London.

Rossum, G.D.v. (2020). 'Clocks, Clock Time and Time Consciousness in the Visual Arts.' *Material Histories of Time: Objects and Practices, 14th–19th Centuries*. Bernasconi, G. and Thürigen, S. (eds.). Walter Gruyter, Berlin, pp. 71–88.

Saliba, G. (2011). *Islamic Science and the Making of the European Renaissance*. MIT Press, Massachusetts.

Salomons, D.L. (2021). *Breguet 1747–1823*. Reprint by Alpha Editions.

Sandoz, C. (1904). *Les Horloges et les Maîtres Horologeurs à Besançon; du XVᵉ Siècle a la Révolution Française.* J. Millot et Cie, Besançon.

Scarsbrick, D. (1994). *Jewellery in Britain 1066–1837: A Documentary, Social, Literary and Artistic Survey.* Michael Russell (Publishing) Ltd, Norwich.

Scott, R.F. (1911–12). *Scott's Last Expedition.* (1941 ed.) John Murray, London.

Seneca, L.A. (c. 49 AD). *On the Shortness of Life.* Penguin, London.

Shaw, M. (2011). *Time and the French Revolution.* The Boydell Press, Suffolk.

Snir, A., Nadel, D., Groman-Yaroslavski, I., Melamed, Y., Sternberg, M., Bar-Yosef, O. et al. (2015). 'The Origin of Cultivation and Proto-Weeds, Long Before Neolithic Farming'. *PLoS ONE*, 10 (7). Available at: https://www.sciencedaily.com/releases/2015/07/150722144709.htm (Accessed 10 August 2020).

Sobel, D. and Andrewes, W.J.H. (1995). *The Illustrated Longitude: The True Story of a Lone Genius Who Solved the Greatest Scientific Problem of His Time.* Fourth Estate, London.

Sobel, D. (2005). *Longitude: The True Story of a Lone Genius Who Solved the Greatest Scientific Problem of His Time.* Walker & Company, New York.

Stadlen, N. (2004). *What Mothers Do (Especially When It Looks Like Nothing).* Piatkus Books, London.

Steiner, S. (2012). 'Top Five Regrets of the Dying'. *Guardian*, 1 February. Available at: https://www.theguardian.com/lifeand-style/2012/feb/01/top-five-regrets-of-the-dying (Accessed 23 July 2020).

Stern, T. (2015). 'Time for Shakespeare: Hourglasses, Sundials, Clocks, and Early Modern Theatre'. *Journal of the British Academy*, vol. 3, 1–33 (19 March).

Stubberu, S.C.; Kramer, K A. and Stubberud, A.R. (2017). 'Image Navigation Using a Tracking-Based Approach'. *Advances in Science, Technology and Engineering Systems Journal*, 2 (3), pp. 1478–86.

Sullivan, W. (1972). 'The Einstein Papers. A Man of Many Parts'.

New York Times, 29 March. Available at: https://www.nytimes.com/1972/03/29/archives/the-einstein-papers-a-man-of-many-parts-the-einstein-papers-man-of.html (Accessed 14 May 2021).

Tann, J. (2015). 'Borrowing Brilliance: Technology Transfer across Sectors in the Early Industrial Revolution'. *International Journal for the History of Engineering and Technology*, 85 (1), pp. 94–114.

Taylor, J. and Prince, S. (2020). 'Temporalities, Ritual, and Drinking in Mass Observation's Worktown'. *The Historical Journal*. Cambridge University Press, pp. 1–22.

Thompson, A. (1842). *Time and Timekeepers*. T. & W. Boone, London.

Thompson, E.P. (1967). 'Time, Work-Discipline, and Industrial Capitalism'. *Past & Present*, 38, (December), pp. 56–97.

Thompson, D. (2007). *Watches in the Ashmolean Museum*. Ashmolean Handbooks. Ashmolean Museum, Oxford.

Thompson, D. (2014). *Watches*. British Museum Press, London.

Thompson, W.I. (2008). *The Time Falling Bodies Take to Light: Mythology, Sexuality and the Origins of Culture*. Digital printed ed. St. Martin's Press, New York.

Unknown Author (2019). 'BBC documentary examines the deep scars left from Dundee Timex closure, 26 years on'. *Evening Telegraph*, 15 October. Available at: https://www.eveningtele-graph.co.uk/fp/bbc-documentary-examines-the-deep-scars-left-from-dundee-timex-closure-26-years-on (Accessed 14 May 2021).

Various (1967). *Pioneers of Precision Timekeeping*. A symposium published by the Antiquarian Horological Society as Monograph No. 3.

Verhoeven, G. (2020). 'Time Technologies'. *Material Histories of Time: Objects and Practices, 14th–19th Centuries*. Bernasconi, G. and Thürigen, S. (eds). Walter de Gruyter, Berlin, pp. 103–115.

Wadley, L. (2020). *Early Humans in South Africa Used Grass to Create Bedding, 200,000 years ago*. YouTube Video. Available at: https://www.youtube.com/watch?v=AzUui4eZI2I (Accessed 8 November 2020).

Walker, R. (2013). *Blacks and Science Volume One: Ancient Egyptian*

Contributions to Science and Technology and the Mysterious Sciences of the Great Pyramid. Reklaw Education Ltd, London.

Weiss, A. (2010). 'Why Mexicans celebrate the Day of the Dead.' *Guardian*, 2 November. Available at: https://www.theguardian. com/commentisfree/belief/2010/nov/02/mexican-celebrate-day-of-dead (Accessed 2 September 2020).

Weiss, L. (1982). *Watch-making in England, 1760–1820*. Robert Hale Ltd, London.

Wesolowski, Z.M. (1996). *A Concise Guide to Military Timepieces 1880–1990*. Reprint. The Crowood Press, Wiltshire.

Whitehouse, D. (2003). '"Oldest sky chart" found'. BBC, 21 January. Available at: http://news.bbc.co.uk/1/hi/sci/tech/2679675.stm (Accessed 12 June 2020).

Wilkinson, C. (2009). *British Logbooks in UK Archives 17th–19th Centuries. A Survey of the Range, Selection and Suitability of British Logbooks and Related Documents for Climatic Research* [online].

Wragg Sykes, R. (2020). *Kindred: Neanderthal Life, Love, Death and Art*. Bloomsbury Sigma, London.

Yazid, M.; Akmal, A.; Salleh, M.; Fahmi, M.; Ruskam, A. (2014). 'The Mechanical Engineer: Abu'l –'Izz Badi'u'z – Zaman Ismail ibnu'r – Razzaz al Jazari' (seminar on Religion and Science: Muslim Contributions Semester 1 2014/2015, 9 December, Skudai, Johor, Malaysia.)

Yoshihara, N. (1985). '"Cheap Chic" Timekeepers: Swatch Watches Offer Many Scents, Patterns'. *Los Angeles Times*, 21 June. Available at: https://www.latimes.com/archives/la-xpm-1985-06-21-fi-11660-story.html (Accessed 14 May 2021).

Zaimeche, S. (2005). *Toledo*. Foundation for Science Technology and Civilisation. June 2005. Pub. ID 4092.

Zaslavsky, C. (1992). 'Women as the First Mathematicians'. *International Study Group on Ethnomathematics Newsletter*, 7 (1), January.

Zaslavsky, C. (1999). *Africa Counts: Number and Pattern in African Cultures*. 3rd ed. Lawrence Hill Books, Chicago.

Further Resources

O bject-led histories are only possible if you can physically see and examine the object of your study. This book would not have been possible without support from some of the many horological collections in museums and art galleries around the world.

If you'd like to see some of the timekeepers I describe in this book, and other examples like them, the following is a list of museums with collections of watches and clocks on public view. Some horological collections are dispersed within wider exhibitions of art and design, and might require a bit of leg work to see them all. Some of these museums are small and open part-time; some offer guided tours and behind-the-scenes visits if you book in advance. I recommend contacting them before your visit to try to find out what objects are currently on display or whether a curator has availability to show you things that aren't!

EUROPE

United Kingdom
Bury St Edmunds: Moyse's Hall Museum
Coventry: Coventry Watch Museum
London: Clockmakers' Company Collection, Science Museum
London: British Museum
London: Royal Observatory

London: Wallace Collection
London: Victoria & Albert Museum
Newark: Museum of Timekeeping in Newark
Oxford: Ashmolean Museum
Oxford: History of Science Museum

Austria
Karlstein: Uhrenmuseum
Vienna: Uhrenmuseum of the Wien Museum

Belgium
Mechelen: Horlogeriemuseum

Denmark
Aarhus, Den Gamle By: The Danish Museum of Clocks and Watches

Finland
Espoo: Finnish Museum of Horology

France
Besançon: Musée du Temps
Cluses: Musée de l'Horlogerie et du Décolletage
Paris: Conservatoire National des Arts et Métiers
Paris: Musée des Arts et Métiers
Paris: Breguet Museum
Saint-Nicolas d'Aliermont: Musée de l'Horlogerie

Germany
Albstadt: Philipp-Matthäus-Hahn-Museum
Furtwangen: Deutsche Uhrenmuseum
Glashütte: Deutsches Uhrenmuseum
Harz, Bad Grund: Uhrenmuseum
Nuremberg: Uhrensammlung Karl Gebhardt
Pforzheimer: Technisches Museum der Pforzheimer, Schmuck und
    Uhrenindustrie
Schramberg: Junghans Terrassenbau Museum

Italy
Bardino Nuovo: Museo dell'Orologio di Tovo S. Giacomo
Milan: Museo Nazionale della Scienza e della Tecnologia Leonardo
   da Vinci

Netherlands
Franeker: Eise Eisinga Planetarium
Joure: Museum Joure
Zaandam: Museum Zaanse Tidj

Romania
St Ploiesti Prahova: Nicolae Simache Clock Museum

Russia
Moscow: Museum Collection
Siberia, Angarsk: Angarsk Clock Museum
St Petersburg: The State Hermitage Museum

Spain
Madrid: Museo del Reloj Antiguo

Switzerland
Basel: Haus zum Kirschgarten – Historisches Museum
Fleurier: L.U.CUEM – Traces of Time
Geneva: Musée d'Art et d'Histoire
Geneva: Patek Philippe Museum
La Chaux-de-Fonds: Musée International d'Horlogerie (MIH)
Le Locle: Château des Monts
Vallée de Joux: Espace Horloger
Zurich: Beyer Clock and Watch Museum

AFRICA

South Africa
Kimberley: McGregor Museum

ASIA

China
Beijing: The Palace Museum, Forbidden City
Macau: Macau Timepiece Museum
Yantai: Polaris Heritage Museum of Clock and Watches

Japan
Hiroshima: The Hiroshima Peace Memorial Museum
Nagano: Gishodo
Tokyo: National Museum of Nature and Science
Tokyo: The Seiko Museum
Tokyo: Daimyo Clock Museum

Thailand
Bangkok: Antique Clock Museum

AUSTRALASIA

Australia
Melbourne: Museums Victoria

New Zealand
Whangarei: Claphams Clock Museum

MIDDLE EAST

Israel
Jerusalem: The Salomons Collection, Meyer Museum of Islamic Art.

Turkey
Istanbul: Topkapı Palace

NORTH AMERICA

Canada
Alberta, Peace River: The Alberta Museum of Chinese Horology
Ontario, Deep River: The Canadian Clock Museum

Mexico
Mexico City: Museo del Tiempo
Puebla: Museo de Relojeria

USA
California, San Francisco: The Interval – The Long Now Foundation
Connecticut, Bristol: American Watch & Clock Museum
District of Columbia, Washington: National Air and Space Museum
Illinois, Evanston: Halim Time & Glass Museum
Maryland, Baltimore: B&O Railroad Museum
Massachusetts, North Grafton: The Willard House and Clock Museum
Massachusetts, Waltham: Charles River Museum
New York, New York: Metropolitan Museum of Art
New York, New York: The Frick Collection
Ohio, Harrison: Orville R. Hagans History of Time Museum (AWCI)
Pennsylvania, Columbia: National Watch & Clock Museum (NAWCC)
Pennsylvania, Philadelphia: Philadelphia Museum of Art
Texas, Lockhart: Southwest Museum of Clocks and Watches

SOUTH AMERICA

Brazil

São Paulo: Museu do Relógio (Professor Dimas de Melo Pimenta)

There are many more museums with horological collections than those I've listed here.

Notes

A full list of works referenced can be found in the Bibliography on p. 221.

A Backward-facing Foreword

p. xiv 'In fact, the word *time* . . .' According to the BBC, cited by Hom, A. (2020). The top ten nouns include two further time words: 'year' and 'day'.

Chapter 1: Facing the Sun

p. 4   'Archaeologists found more . . .' Wadley, L. (2020).

p. 5   'If our ancestors rotated . . .' Walker, R. (2013), p. 89.

p. 5   'A recent study found no . . .' Helfrich-Förster, C.; Monecke, S.; Spiousas, I.; Hovestadt, T.; Mitesser, O. and Wehr, T.A. (2021).

p. 7   'In other words, they too . . .' Häfker, N.S.; Meyer, B.; Last, K.S.; Pond, D.W.; Hüppe, L.; Teschke, M. (2017), p. 2194.

p. 7   'If anything, our ability to . . .' Popova, M. (2013).

p. 7   'The Lebombo Bone suggests . . .' The Pradelles hyena bone is another example of a bone that has been marked with regular parallel incisions that look completely unlike anything you would expect from butchery, and dates to a similar time as the Lebombo Bone. Where it differs is that the Pradelles bone was carved by our cousins, the Neanderthals. The marks may well be decoration or are possibly evidence of some level of numeracy. See Wragg Sykes, R. (2020), p. 254.

p. 7 'In the Democratic Republic . . .' The Ishango Bone currently resides on permanent public display at the Royal Belgian Institute of Natural Sciences in Brussels, Belgium.

p. 7 'As the philosopher William Irwin Thompson . . .' Thompson, W.I. (2008), p. 95.

p. 8 'This site is the earliest . . .' Snir, A.; Nadel, D.; Groman-Yaroslavski, I.; Melamed, Y.; Sternberg, M.; Bar-Yosef, O.; et al. (2015).

p. 9 'This is how the Kenyan-born . . .' Zaslavsky, C. (1999), p. 62.

p. 9 'For some coastal Aboriginal . . .' Wragg Sykes, R. (2020), pp. 278–9.

p. 10 'For them, a meaningful . . .' Rameka, L. (2016), p. 387.

p. 11 'The Sumerians, the first . . .' Zaslavsky, C. (1999), p. 23.

p. 11 'About 5,000 years ago . . .' Locklyer, N.J. (2006), p. 110. Dedicated astronomers that they were, the Egyptians may even have pioneered heliocentric theory – the theory that the Earth and other planets revolve around the sun – though for centuries Ptolemy of Alexandria's Earth-centred model of the second century CE held sway.

p. 12 'In the hours of darkness . . .' Ibid., p. 343.

p. 12 'Astronomers identified . . .' Walker, R. (2013), p. 16.

p. 13 'They also knew of the planets . . .' Ibid., pp. 18–19.

p. 14 'In ninth-century England . . .' The description was made in the accounts of King Alfred's royal biographer, Bishop Asser, which was written in 893 CE.

p. 14 'In his Documenti d'Amore . . .' Balmer, R.T. (1978), p. 616.

p. 14 'In the late fifteenth century . . .' May, W.E. (1973), p. 110.

Chapter 2: Ingenious Devices

p. 20 'In ancient Mesopotamia . . .' Masood, E. (2009), p. 163.

p. 20 'We don't know exactly . . .' Ibid., p. 74.

p. 21 'Al-Zarqali managed . . .' Zaimeche, S. (2005), p. 10.

p. 22   'Al-Zarqali's clock remained . . .' Masood, E. (2009), p. 74.

p. 22   'Nearly 4,000 miles east . . .' Foulkes, N. (2019), p. 64.

p. 22   'The brief for Su Song's . . .' Morus, I.W. (2017), p. 108, as cited by Foulkes, N. (2019), p. 65.

p. 22   'With its bronze . . .' Foulkes, N. (2019), p. 65. Sadly, Su Song's clock has been lost in history. It was taken during the Tatar invasion of China in 1127 but – similar to the fate of Al-Zarqali's clock in Toledo – the invaders' scholars were unable to put it together and make it run again. Today, the closest example we have is a fully functioning scale replica that stands outside the Gishodo Suwako Watch and Clock Museum in the Suwa area, one of the great horological centres of Japan, and a short drive from the famous watch company Seiko.

p. 23   'They include automata . . .' Yazid, M.; Akmal, A.; Salleh, M.; Fahmi, M.; Ruskam, A. (2014).

p. 23   'A mahout, or elephant trainer . . .' Masood, E. (2009), p. 163.

p. 26   'The Elizabethan dramatist . . .' Stern, T. (2015), p. 18, citing a seventeenth-century manuscript compiled by a Richard Smith and quoted in Bedini, Doggett & Quinones (1986), p. 65.

p. 26   'In late medieval . . .' Glennie, P. & Thrift, N. (2009), p. 24.

p. 26   'In an age when . . .' Ibid.

p. 28   'Medieval church clocks . . .' Oestmann, G. (2020), p. 42.

p. 28   'Devices like lunar volvelles . . .' An example of a lunar volvelle can be found at Bodleian Libraries, University of Oxford, accession number MS. Savile 39, fol. 7r, https://www.cabinet.ox.ac.uk/lunar-tool#/media=8135 (accessed 19 April 2021).

p. 29   'Astronomers drove clocks . . .' Baker, A. (2012), p. 16.

p. 29   'The regimented nature . . .' Álvarez, V.P. (2015), p. 64.

p. 29   'The story goes . . .' Johnson, S. (2014), p. 137.

p. 31   'Over the course . . .' Álvarez, V.P. (2015), p. 65.

p. 34   'In 1511, the scholar . . .' Lester, K. & Oerke, B. V. (2004), p. 376. Although the claim of 'forty-eight hours' is dubious as watches of this era rarely run longer than a day.

Chapter 3: Tempus Fugit

p. 36   'On the forehead . . .' Horace: '*Pallida mors æquo pulsat pede pauperum tabernas Regumque turres*' ('Pale Death, with impartial foot, knocks at the cottages of the poor and the palaces of kings').

p. 38   'At the end of . . .' Thompson, A. (1842), pp. 53–54.

p. 38   'But in the early 1980s . . .' Jagger found evidence that a skull watch – presumed to have been Mary's – was known to have existed in Salisbury in 1822, and this, in turn, could have led to the creation of a further two in the nineteenth century. A letter written in 1863 by Queen Victoria's clockmaker in Scotland describes a 'death's head watch, formerly belonging to Mary Queen of Scots', which had come into the family of Sir John Dick Lauder through Catherine Seton, sister of Mary Seton, 'to whom the unfortunate Mary gave it before her execution'. Yet an 1895 transactions ledger from the St Paul's Ecclesiological Society lists the sale of 'a devotional watch, with a very sweet tones bell, on which the hours are struck, set in a case in the form of a skull, covered with engravings . . . It very closely resembles the watch belonging to Sir Thomas W. Dick-Lauder . . . which was by him stated to be one of the twelve presented by Mary to her favourite ladies of honour'.

p. 38   'It would have had . . .' An inventory made while Mary was at Holyrood in 1562 lists an impressive range of items that included sixty gowns – many heavily embroidered – in her favourite colour, white, as well as jet black, crimson red, and orange with silver detail. There are cloths of gold, silver, velvet, satin and silk waiting to be made up into clothes. She had fourteen cloaks, mantles (similar to a cloak only without sleeves) of purple velvet and ermine, and thirty-four vasquines, or corsets.

p. 40   'Her letters and the quotes . . .' The quote '*En ma Fin gît mon Commencement*' ('In my end is my beginning') was said to have been embroidered by Mary during her imprisonment.

p. 40   'When they stripped . . .' Fraser, A. (2018), p. 669.

p. 41   'In a portrait of King Henry . . .' There are a number of Holbein portraits that decisively depict the watches and clocks of his wealthy patrons, including those of the French ambassador Charles de Solier (1480–1552), the Swiss merchant Jörg Gisze (1497–1562), and the group portrait of lawyer Sir Thomas More (1478–1535) and his family.

p. 41   'The list, drawn up . . .' Cummings, G. (2010), p. 14.

p. 41   'One was more spherical . . .' Incense was important not only to help sixteenth-century Europeans with refined noses navigate the olfactory challenges of living in a city without a proper sewerage system, but it was also believed to ward off diseases like the plague.

p. 42   'After carefully removing . . .' As things stand, we still don't know who buried the hoard, when, or why. The treasure's custodian, Hazel Forsyth, has made the pursuit of these questions into a significant body of work. Her theories are that it could have been a goldsmith heading off to fight in the Civil War, which started in 1642, from which he never returned. It's also possible that he was fleeing that conflict to seek safety abroad. It's likely it belonged to a Jacobean goldsmith who had been resident at that site, but whoever he was, it is clear he wasn't around for long enough to reclaim his buried treasure.

p. 46   'Over the next two decades . . .' For atrocities on such a large scale that happened nearly five hundred years ago, with little attempt to officially document them, it is understandable that estimates for the number of dead and displaced vary greatly between sources. For these figures I have referenced Murdoch, T.V. (1985), p. 32.

p. 49   'Flamboyant dress was . . .' Ribero, A. (2003), p. 65.

p. 49   'Even the powdered wigs . . .' Ibid., p. 73.

p. 49   'Puritans dressed modestly . . .' Watches were already such invaluable tools that even strict Puritans couldn't give them up. There is a watch in the collection of the British Museum with provenance linking it, albeit somewhat questionably, to Cromwell, while another example that also has claims to once serving the Puritan leader appeared at auction in Carlisle, Cumbria, in 2019.

p. 50   'They believed that to prosper . . .' The Huguenots in particular were renowned for their industrious working habits. In 1708 Edward Wortley described to the House of Commons how work for them was 'the practical exercise of a calling appointed by God'. Watchmaker David Bouguet was a devout Protestant – he served as an elder at the French Church on Threadneedle Street four times – and would likely have held the view that not to apply his God-given gifts to the best of his ability would have been sinful. To those of Calvinist faith, time is an example of God's plan manifested in nature and its observance was a matter of religious importance.

p. 50   'In the Puritan day . . .' Rossum, G.D.v. (2020), p. 85.

p. 50   'It was even argued . . .' Richardson, S. (1734) as cited by Rossum, G.D.v. (2020), p. 85.

p. 50   '"Time being man's opportunity . . ."' Baxter, R. (1673).

p. 51   'As his reign progressed . . .' Fraser, A. (1979) as cited by Rossum, G.D.v. (2020), p. 74.

p. 52   'In the years that followed . . .' Murdoch, T.V. (1985), p. 51.

p. 52   'While the majority . . .' Ibid.

p. 52   'These Huguenot migrants . . .' Thompson, W.I. (2008), p. 40.

Chapter 4: The Golden Age

p. 57   '"So home and late . . ."' Samuel Pepys' diary entry from Tuesday 22 August 1665, https://www.pepysdiary.com/diary/1665/08/22/.

p. 59   'When I take apart . . .' House of Commons (1818), p. 4. Any object which has required significant human effort to create becomes authored and will exhibit some degree of unique personalisation both in the subtleties of the finishing and the obvious fingerprints of the craftsman (such as signatures and maker's marks). With a trained eye, these marks can be read like a text.

p. 60 'Their watches were charged . . .' Cummings, N. & Gráda, C.Ó. (2019), pp. 11–12.

p. 65 'In 1777, watchmaker . . .' Dickinson, H.W. (1937), p. 96; Tann, J. (2015) as cited by Cummings, N. & Gráda, C.Ó. (2019), p. 19.

p. 65 'According to a 1798 . . .' Thompson, E.P. (1967), p. 65.

p. 66 'Some of the stars . . .' Stubberu, S.C.; Kramer, K.A.; Stubberud, A.R. (2017), p. 1478.

p. 67 'He was also able . . .' Abulafia (2019), pp. 17, 812–13.

p. 67 'And nor, in his opinion . . .' Sobel, D. (1995), p. 52.

p. 71 'By the time she returned . . .' Robert FitzRoy as cited in Barrie, D. (2014), p. 227.

p. 71 'They are liable . . .' Henry Raper as cited in Barrie, D. (2014), p. 89.

p. 71 'Most early-modern navigational . . .' Baker, A. (2012), p. 15.

p. 72 'Shortly after, a key . . .' Baker, A. (2012), pp. 23–24.

p. 72 '"Trim took a fancy . . ."' National Maritime Museum, Flinders' Papers FLI/11 as cited in Barrie, D. (2014), p. 204.

p. 74 'And although the Board of Longitude . . .' Wilkinson, C. (2009), p. 37.

p. 74 'In 1802, thirty years . . .' Rodger, N.A. (2005), pp. 382–3 as cited in Barrie, D. (2014), p. 115.

p. 75 'It is the descendant . . .' Good, R. (1965), p. 44.

Chapter 5: Forging Time

p. 81 'All of them, like Wilter's . . .' Although Mudge had introduced his lever escapement in 1767, it was to the average pocket watch of the day as the Zenith Defy (currently proclaiming itself to be the most accurate mechanical watch in the market) is to a simple Timex. The verge escapement was still the standard.

p. 82 'Watches were complicated . . .' Chapuis & Jaquet (1970), pp. 80–82.

p. 82 'One worker in the north . . .' Heaton, H. (1920), pp. 306–11 as cited by Cummings, N. & Gráda, C.Ó. (2019), p. 6.

p. 83   'One observer noted . . .' Ganev, R. (2009), pp. 110–11.

p. 83   'Some local manufacturers . . .'Cummings, N. & Gráda, C.Ó. (2019), p. 6.

p. 83   'Artisan culture was almost . . .' Clarke, A. (1995) as cited by Ganev, R. (2009), p. 5.

p. 83   'The number of women serving . . .' Erickson, A.L., p. 2.

p. 83   'One ongoing study . . .' The list is frequently updated and the latest figure can be viewed at https://www.ahsoc.org/resources/women-and-horology/.

p. 87   'As a result of *établissage* . . .' Landes, D. (1983), p. 442.

p. 88   'Renowned father–son watchmakers . . .' Neal, J.A. (1999), p. 109.

p. 89   'He justified it . . .' Pitt introduced a plethora of taxes over this time, including Income Tax, to compensate for the financial burden of the French Revolutionary and inevitable Napoleonic Wars.

p. 90   'The closer Moll gets . . .' Rossum, G.D.v. (2020), p. 78.

p. 90   'The proliferation of watches . . .' Ibid., p. 73.

p. 90   'Timepieces were highly prized . . .' Ibid., p. 86.

p. 90   'Records from the Old Bailey . . .' Styles (2007) as cited by Verhoeven, G. (2020), p. 111.

p. 91   'Time awareness was slowly . . .' Verhoeven, G. (2020), p. 105. Increases in clock ownership also contributed to a rise in time awareness. In 1675, barely 11 per cent of London households had a clock; thirty years later the figure stood at 57 per cent. In the 1770s, more than 10 per cent of all cases that went through the Old Bailey related to the theft of clocks. While this was going on, their average prices decreased, falling by as much as 75 per cent over the century. Nonetheless, over half the reports of clock theft from the late eighteenth and early nineteenth centuries were from affluent owners, despite their being a minority of the population.

p. 92   '"[He] introduced the making . . ."' House of Commons (1812), p. 67.

Chapter 6: Revolution Time

p. 97   'The barrister and horology buff . . .' Salomons, D.L. (1921), p. 5.

p. 98   'By moving the dial . . .' Credit for this analogy must be attributed to the wonderfully talented watch restorer and rose engine turner Seth Kennedy.

p. 100   'It was Breguet's stepfather . . .' Researching Breguet's personal life is challenging. He was a celebrity in his day, and has been commonly written about ever since, with little by way of hard facts for anything outside his watchmaking endeavours. There are books about his life that make no mention of his family at all; even dates for the completion of his works can vary. I have done my best to tread a middle ground.

p. 100   'He lost his father at . . .' Depending on the source, Breguet's age at the loss of his father has been quoted as ten, eleven or twelve years old.

p. 103   '"This devil, Breguet . . ."' Breguet, C. (1962), p. 5.

p. 103   'Breguet "had the power..."' Ibid., p. 6.

p. 103   'In the months that followed . . .' Also referred to as 'the Terror' and 'the Reign of Terror'.

p. 103   'Although the most famous . . .' Ironically, the guillotine had been invented as a more humane method of execution by a medical doctor opposed to execution, Dr Joseph-Ignace Guillotin. The machine was based on previous sliding axe designs used in Italy and Scotland, improved and refined to ensure a swift and clean despatch.

p. 103   'The accounts of the survivors . . .' These accounts and many more like them can be found in Anon (1772).

p. 104   'In April 1793 . . .' Antiquorum (1991).

p. 104   'Two months later . . .' Daniels, G. (2021), p. 6.

p. 105   'So important was this . . .' The watch mentioned went under the hammer at Sotheby's on 14 July 2020, selling for £1,575,000. My description was aided by their cataloguing which can be read in full online at https://www.sotheby's.com/en/buy/

auction/2020/the-collection-of-a-connoisseur/breguet-
retailed-by-recordon-london-a-highly.

p. 106   'As Jamaican philosopher . . .' Mills, C. (2020), p. 301.

p. 106   'Decimal time, as it was known . . .' Shaw (2011).

p. 106   'This new calendar was . . .' To balance leap years, every fourth
          year there would be an additional 'Festival of the Revolution'.

p. 108   'His cover story . . .' Daniels, G. (2021), p. 7.

p. 108   'Breguet agreed . . .' Ibid., p. 9.

p. 109   'In his last years . . .' Salomons, D.L. (2019), pp. 11–12.

p. 110   'Stendhal declared Breguet's watches . . .' Breguet, C. (1962),
          p. 10.

Chapter 7: Working to the Clock

p. 112   'Before industrialisation, Britain . . .' Published in 1967, British
          social historian and political campaigner E.P. Thompson's
          seminal essay 'Time, Work-Discipline and Industrial Capitalism'
          is an exceptional source for discovering how heavily time
          transformed from a force of nature to a force of organisational
          control. If you find this section of interest, I strongly advise
          tracking down the original essay and those it inspired.

p. 113   'This change was less about . . .' Thompson, E.P. (1967), p. 61.
          The idea of a working day and of paying an hourly rate dates
          back to the sixteenth century: see Glennie and Thift (2009),
          p. 220.

p. 114   'In 1757, the Irish statesman . . .' Edmund Burke, cited by Ganev,
          R. (2009), p. 125.

p. 115   'He describes stepping . . .' Myles, J. (1850), p. 12. Chapters in
          the Life of a Dundee Factory Boy, an autobiography, James
          Myles, 1850

p. 116   'Their services became more . . .' Peek, S. (2016). Knocker-
          uppers were such an important part of London life that Charles
          Dickens mentions them in Great Expectations, with Mr Whopsle
          being woken by a knocker-upper. The job was far from short-
          lived and continued until relatively recently: it wasn't until the

1970s that the UK's last knocker-upper hung up their pea shooter.

p. 116 'An account from a mill worker . . .' Alfred, S. K. (1857) quoted by Thompson, E.P. (1967), p. 86.

p. 117 'As one industrialist explained . . .' Rev. J. Clayton's *Friendly Advice to the Poor* (1755), quoted by Thompson, E.P. (1967), p. 83.

p. 117 'In 1770 English cleric . . .' Temple (1739–1796) was from Berwick-upon-Tweed in Northumberland and had been educated at the University of Edinburgh. He published a number of essays expressing his views on religion, power and morality.

p. 117 'In 1772, in an essay . . .' Anon. (1772).

p. 117 The watch industry had . . .' Newman, S. (2010), p. 124.

p. 120 'The year in which you are reading . . .' Mills, C. (2020), p. 300.

p. 121 'They were "grounded in ..."' Mills, C. (2020), p. 308.

p. 121 'Western observers in the 1800s . . .' Thompson, E.P. (1967), pp. 91–92.

p. 122 'These kinds of stereotypes . . .' Hom, A. (2020), p. 210.

p. 122 'Needless to say, it was not . . .' Thompson, E.P. (1967), pp. 56–97.

p. 124 '"when Home we are come . . ."' Collier, M. (1739), pp. 10–11. 'The Woman's Labour: an Epistle to Mr. Stephen Duck; in Answer to his late Poem, called The Thresher's Labour' (1739), pp. 10–11.

p. 124 'A similar eighteenth-century . . .' Ganev, R. (2009), p. 120.

p. 125 'New parents working in the home . . .' Stadlen, N. (2004), p. 86. One of the new mothers Stadlen interviews declares, 'Clock time doesn't mean anything any more.'

p. 125 'If a memory is unclear . . .' Sophie Freeman, 'Parents find time passes more quickly, researchers reveal', *The Times*, 22 February 2021; Popova, M. (2013).

p. 126 'Thousands of watchmakers . . .' House of Commons (1817), p. 15.

p. 126 '"hardly a rag to cover them..."' House of Commons (1817), p. 5.

p. 126 'By the mid-nineteenth century . . .' Hoult, J., 'Prescot Watch-making in the xviii Century', *Transactions of the Historic Society of Lancashire and Cheshire*, LXXVII (1926), p. 42, as cited by Cummings, N. & Gráda, C.Ó. (2019), p. 27.

p. 128   'In 1878 one unnamed . . .' Church, R.A. (1975), p. 625 as cited by Cummings, N. & Gráda, C.Ó. (2019), p. 24.

Chapter 8: The Watch of Action

p. 131   'Railways even shrank . . .' Ramirez, A. (2020), p. 49.

p. 131   'By the 1850s . . .' Ibid., figs 18–19.

p. 132   'It's not until you see . . .' 'Corn Exchange Dual-Time Clock', Atlas Obscura, https://www.atlasobscura.com/places/corn-exchange-dualtime-clock.

p. 133   'Dual time observations . . .' Bartky, I. (1989), p. 26.

p. 134   'He described it as his . . .' Slocum as cited in Barrie, D. (2014), p. 245.

p. 135   'Likewise the black . . .' The dangers of luminous paint were brought to mainstream attention through the tragic harm they caused the Radium Girls, who painted these dials by hand with a fine brush in the 1910s and 1920s, and whose story we will return to in the next chapter.

p. 137   'He even schedules . . .' Scott, R.F. (1911), p. 235.

p. 137   'It's hard to imagine . . .' Ibid., p. 210.

p. 138   'While Britain's scorched . . .' 'South African concentration camps', New Zealand History, https://nzhistory.govt.nz/media/photo/south-african-concentration-camps (accessed 23 November 2022).

p. 139   'At home the wristlet . . .' 'The History of the Nato Watch Strap', A. F. 0210, https://afo210strap.com/the-history-of-the-nato-watch-strap-nato-straps-in-the-great-war-wwi-era/ (accessed 12 January 2023).

p. 139   'In 1893, an advert . . .' Ibid.

p. 139   'Another, from 1901 . . .' Ibid.

p. 141   'At around 700 metres . . .' Geffen (2010).

p. 141   'Within the remaining . . .' Gabardine is a tough and durable tightly woven cloth, typically made from wool or cotton, which was commonly used in the production of uniforms, coats and outdoor wear.

Chapter 9: Accelerated Time

p. 148   'One account from 1900 . . .' Gohl, A. (1977), p. 587 as cited in Glasmeier, A. (2000), p. 142.

p. 148   'In one 1915 edition . . .' Cartoonist M.C. Fisher cited in Cummings, G. (2010), p. 232.

p. 150   'In 1908 Wilsdorf . . .' Dowling, J.M. & Hess, J.P. (2013), p. 11.

p. 154   'By 1926, US manufacturer . . .' Moore, K. (2016), p. 171.

p. 154   'By the end of 1918 . . .' Ibid., p. 25.

p. 155   'Each and every dial . . .' Ibid., p. 9.

p. 155   'On average, women . . .' Ibid., p. 11.

p. 155   'Although adverts asked . . .' Ibid., p. 45.

p. 155   'Radioluminescent paint . . .' Ibid., p. 8.

p. 155   Dial painters had to . . .' Ibid., pp. 7–8.

p. 156   'In Switzerland . . .' Ibid., p. 10.

p. 156   'The "lip, dip, paint" . . .' Ibid., pp. 9–10.

p. 156   'Some of the painters . . .' Ibid., p. 16.

p. 156   'He himself had amputated . . .' Ibid., pp. 18–19.

p. 157   'The body, mistaking . . .' Ibid., p. 111.

p. 157   'Doctors, including those . . .' Ibid., p. 224.

Chapter 10: Man and Machine

p. 170   'The auction house initially . . .' Davidson, H. (2021).

p. 173   '(Adverts for the Accutron . . .)' 'Reinventing Time: The Original Accutron', Hodinkee, https://www.hodinkee.com/articles/reinventing-time-original-bulova-accutron (accessed 23 November 2022).

p. 175   'By the early 1980s . . .' Glasmeier, A. (2000), p. 243.

p. 176   'It was advertised as . . .' Finlay Renwick, 'The Digital Watch Turns 50: A Definitive History', Esquire, 18 November 2020, https://www.esquire.com/uk/watches/a34711480/digital-watch-history/.

p. 177   'In 1985 the *LA Times* . . .' Yoshihara, N. (1985).

p. 178   'By the time it closed . . .' Unknown Author (2019).

p. 178   'By 1993, Hong Kong . . .' *South China Post* (1993), p. 3; Hong Kong Trade and Development Council (1998) as cited in Glasmeier, A. (2000), p. 231.

p. 178   'manufacturers could supply . . .' Glasmeier, A. (2000), p. 233.

p. 178   'One protester described . . .' Unknown Author (2019).

p. 185   'The Deep Space Atomic Clock . . .' 'Deep Space Atomic Clock', NASA, https://www.nasa.gov/mission_pages/tdm/clock/index.html (accessed 23 November 2022).

Chapter 11: Eleventh Hour

p. 195   '"you must not think a man . . ."' Seneca, L.A. (c. 49 AD), p. 11.

Index

Abraj Al Bait Towers 27
Accutron watch 173
al-Jazari, Ismail 23–4
al-Rashid, Harun 20
al-Zarqali 21, 22
Albert, Prince 114
Aldrin, Buzz 171
Alfred the Great, King 14
ancient world 11–14
animal world 3, 5–7
Anker, Conrad 141
Antoinette, Marie 94, 96, 101–2, 103
Apple Watch 183
Archie (dog) x, xi, 2, 3, 87
Armstrong, Neil 171
Around the World in Eighty Days (Verne) 130–1, 134
As You Like It (Shakespeare) 15
Astron watch 173–4
astronomy 28–9
atomic clocks 171–2, 185–6
Aurignacians 7
automatic winding 159

Barberino, Francesco da 14
battery-powered watches 172–3
Baxter, Richard 50–1
Bayly, William 72
Beagle, HMS 71
Beggar's Opera (Gay) 90
Bell, Alexander Graham 131
Berlioz, Louis-Hector 194
Berthaud, Ferdinand 96
Besso, Michele 146
Birmingham x, 1–2, 21–2, 59, 148–9
black polishing 70
Blériot, Louis 131
Blois 45
Blue Hawaii 172
Boer War 138–9, 147, 153
Bohr, Niels 172
Bomford, Charles Powell 165
Bouguet, David 46, 47
Bouguet, Hector 47
Bouguet, Solomon 47
Boulton, Matthew 65
Bourges Cathedral 29

Bowman, Robert Anthony
165–6
Bradbury, Norman 125
Breguet, Abraham-Louis 93–105,
107, 108–10
Breguet, Antoine 101
Breguet, Cécile Marie-Louise
100, 101
Breguet overcoil hairspring 97
British Museum 31, 46, 78–80
Buck, Paul 80–1
Bulova 173
Burke, Edmund 114–15

Calvin, John 46, 48
Campaign watch 138–9
Campbell, Sir Malcolm 161
candle clocks 14
Cartier, Louis 139
Casio 176
Catherine de' Medici 46
cats on ships 72–3
chamfering 70
Charlemagne, King 20
Charles I, King 49
Charles II, King 51
Charlotte, Queen 63, 101
Cheapside Hoard 42–4
China 12, 22–3
Chioggia 25
Christchurch Fusee Chain Gang
117–18
Christian Directory, The (Baxter)
50–1
chronemics 120–1

Chung, Cheryl 177
Clarke, Henry 92
Clockmakers' Company 60, 82,
83, 85
clocks
in art and literature 90
atomic clocks 171–2, 185–6
candle clocks 14
in eighteenth century Europe
56
hydromechanical 20–4
in Medieval Europe 24–6,
27–9
in Renaissance Europe 29–32
and space exploration 185–6
turret clocks 26–7, 34
water clocks 13, 14
clockwise movement 11
Cockläus, Johannes 34
Coleman, Bessie 130
Collier, Mary 124
colonialism 120–1
Columbus, Christopher 14
Cooke, Oliver 81
Corday, Charlotte 104
Count of Monte Cristo, The
(Dumas) 110
Cox, Robert Harvey 118
Cromwell, Oliver 48–9, 51
Cromwell, Richard 51
Curie, Marie 154, 173
Curie, Pierre 154, 173

da Vinci, Leonardo 33
Daniels, George 130, 190

Darwin, Charles 71
Davis, Alfred 148–9, 160
Deep Space Atomic Clock 185–6
Defoe, Daniel 90
Dekker, Thomas 26
Dennison, Aaron Lufkin 127, 148–9
Dickens, Charles 111
digital watches 175–7
Diller, Na'aman 94
Documenti d'Amore (Barberino) 14
dog's internal clock 6
Duboule, Jean-Baptiste 48
Dufek, George J. 84
Dumas, Alexander 110
Dürer, Albrecht 33
Dutch forgery watches 76–8, 81, 85–8, 89–90

Earhart, Amelia 140
Eddington, Arthur 152
Edward VI, King 39
Efron-Gabai, Hila 93–4
Egypt, Ancient 11, 12–13
Einstein, Albert 145–6, 164
Elizabeth I, Queen 41
Ellicott, John 62
Encyclopédie (Diderot) 96
établissage system 86–7, 127
Eugène Onegin (Pushkin) 109
Europe
    Early Modern 41–52
    eighteenth century 55–6, 58–62, 80–1
    Medieval 20–1, 24–6, 27–9

Renaissance 29–35
seventeenth century 57–8

Ferdinand, King 60–1
First World War 152–4
Flinders, Matthew 72–3
'Floozie in the Jacuzzi' 21
Floyer, Sir John 65
Francis II, King 39–40
Frayn, Francis Edward 165–8
French Revolution 103–7
Frisius, Gemma 15
fusee chains 33–4, 80–1, 117–18
Fusoris, Jean 29

Galileo 29–30
Gay, John 90
George III, King 63, 101, 105, 109
Glaser, Georg 33
Gleitze, Mercedes 161
gnomons 11, 12
Goebbels, Joseph 34
Goldsmiths' Company 48
Gooch, William 73
Graham, George 60, 61, 68
Greece, Ancient 13
Greenwich Mean Time 133–4
Guise, Duke of 46
Gulliver's Travels (Swift) 53
Gutenberg, Johannes 33

H4 (marine chronometer) 68–9
H5 (marine chronometer) 69–70

hairsprings 58, 59
Hall, Edward T. 120
Halley, Edmond 67
Hamilton Pulsar watch 175–6
Hamilton Ventura watch 172–3
Hard Times (Dickens) 111
Harlan, Walter 34
Harlot's Progress, The (Hogarth)
    90
Harrison, John 68–70, 72, 73–4,
    75
Harrison, William 69
Hasson, Rachel 93
Hayek, Nicolas 177
Henlein, Peter 32–5
Henry II, King 46
Henry IV, King 52
Henry VIII, King 41
Henshaw, Alex 162
Heyden, Gaspard van der 15
Hill, Thomas 189
Hillier, Thomas 91
Hiroshima Peace Memorial Park
    170
Hogarth, William 90
Holbein the Younger, Hans 41
Hom, Andrew 122
Homer 67
Hooke, Robert 58
Hour Angle watch 140–1
hourglasses 14–15
Hughes, Mollie 136
Hugo, Victor 93, 110
Huguenots 46–8, 52
Hume, David 57

Huygens, Christiaan 30, 57
hydromechanical clocks 20–4

Imeson, Gerald 163–4
industrialisation 111–20, 124–9
Ingersoll Watch Company 127–8,
    134
internal clocks 5–7
Ishango Bone 7
Islamic world 23–4

Jagger, Cedric 38
James I, King 48

Kahan, Elli 93
Kneebone, Roger 191
knocker-uppers 116

Larcum, Kendall 73
Larkin, Philip 119
Lawrence, George Fabian 42–3
Lebombo Bone 4–5
Lenin, Vladimir 152
Les Chansons (Hugo) 93, 110
lever escapement 63–5
Lichtenau Castle 34
*Life and Opinions of Tristram
    Shandy, Gentleman, The*
    (Sterne) 57
Lindbergh, Charles 140, 167
Locke, John 57
London in watchmaking 55,
    58–62
Longines 140, 167
longitude 66–75

Louis XVI, King 95, 100
luminous dials 154–8
lunar cycles 5

Maggia, Mollie 157
mainsprings 30–1
Mallory, George 141–2, 143–4, 148
Mappin & Webb 138
Marat, Jean-Paul 104
Marie, Joseph-François 100
marine chronometers 68–75
Mary, Queen of Scots 38–41, 46
Mayotte, Lanny 177
Mbiti, John S. 9
Mebert/Maubert family 47
menstrual cycle 5
Meucci, Antonio 131
migration patterns 3, 5
Mistry, Dhruva 21
Moore, Kate 155
Morris, Liz 84
Morse, Samuel 131
Mount Everest 141–3
Movado 167
movement holders 53–4
Mudge, Thomas 60–2, 63–5, 75
Museum of Timekeeping (Upton Hall) 134–5
Mutt and Jeff 148
Myles, James 115, 116–17

Nahmias, Samuel 93
Newton, Isaac 57, 67–8

Norwich Cathedral 28
Nuremberg 33
Nuremberg Egg, The (Harlan) 34

Odyssey (Homer) 67
Omega 171
On the Shortness of Life (Seneca) 195
Oyster watch 161

pedometers 65
pendulums 29–30, 57–8
Pepys, Samuel 57, 58
perpetual calendars 61
Perry Barr 1
pilot watches 139–41
Pisa Cathedral 29–30
Pitt, William 89
Plato 13
Powys, Caroline Lybbe 63
prehistory 4–5, 7, 8–10
Presley, Elvis 172
pulse watches 65
Pushkin, Alexander 109

quartz watches 173–5

Rabi, Isidor 172
Radium Girls, The (Moore) 155
Rake's Progress, A (Hogarth) 90
Raper, Henry 71
Rebberg watch 151, 152, 153, 184–5
Richard of Wallingford 29
ring dials 15

River, The (Mistry) 21
Robert the Bruce 38
Rolex 150, 151, 152, 160–1,
   162–4, 171
Romieu, Isaac 47
Roosevelt, Theodore 128
Rudolf II, Emperor 31

St Bartholomew's Day Massacre
   46
Salisbury Cathedral 25
Salomons, Sir David Lionel 97
sandglasses 14–15
Santos-Dumont, Alberto 139
Santos pilot watch 139
Sasaki, Kazunari 173
Sassoon, Siegfried 145
Schlottheim, Hans 31
Scott, Robert Falcon 135–7
Sea of Galilee 8
Seag, Morgan 84
Second World War 162–4,
   165–71
Seiko 173–5, 176
Seneca 195
Seton, Mary 38
Shakespeare, William 15
shock settings 102–3
Shovell, Sir Cloudesley 66, 183
skull watches 36–41
Slocum, Joshua 134
smartwatches 183
Smith of Derby 27
Smith, Elinor 140
space exploration 171, 185–6

stars 7, 11, 12–13
Su Song 22–3
Sumeria 11
solar cycles 11–12
standardisation of time 132–4
Sterne, Laurence 57
Stonehenge 11–12
Strauss, Levi 147–8
Struthers, Craig 19, 193
   love of Rebberg watches 151,
      152, 153
   and luminous dials 158
   move to Staffordshire 2
   and Project 248 189–90
   self-employment 119–20,
      150–1
   watchmaking workshop
      ix–xii
   work with watches 44, 96,
      129, 159, 169, 180, 181–3,
      185
Struthers, Rebecca
   as cataloguer for auction house
      76
   collaborative work 44–5
   journey to Mount Everest
      142–3
   love of digital watches 175–6
   and passage of seasons 1–4
   research xiii, 78–80
   self-employment 119–20,
      150–1
   on silversmithing and jewellery
      course 17–19
   stubbornness of 130–1

training as watchmaker 19–20,
53–5, 80–1
watch repair process 197–206
watchmaking workshop ix–xii,
111
work with mechanical watches
180–3, 189–92
subscription watches 108–9
sundials 11–12, 14, 15, 186
Swatch watches 177–8
Swift, Jonathan 53
Switzerland
as centre of watchmaking
87–8
decline of watchmaking in
174, 175
and Second World War 162
and Swatch watches 177–8

Talleyrand-Périgord, Charles
Maurice de 103
Taosi 12
Tattet, Joseph 100
Temple, William 117
Thompson, David 80, 81
Thompson, E.P. 115, 120
Thompson, William Irwin 7
Thurlby, Paul 182
Tiananmen Square massacre
169–70
time
atomic time 172
and colonialism 120–1
and death 39
eighteenth century views of 57

and Einstein's theory of
relativity 145–6
in French Revolution 106–8
and industrialisation 111–20,
124–5
in prehistory 8–10
and Protestantism 50–1
standardisation of 132–4
and watches xiii–xiv, 91
'Time, Work-Discipline and
Industrial Capitalism'
(Thompson) 115
Timex strike 178–9
tolerance 53–4
Tompion, Thomas 61
tourbillon 105
trench watches 152–4
Trevithick, Richard
Trim (cat) 72–3
Tupaia 67
Turner, Laura 81
turret clocks 26–7, 34

verge escapement 24–6, 63–4, 81
Valley of the Kings 12
Vancouver, George 73
Vanity Fair (Thackeray) 110
Verne, Jules 110, 130–1, 134
View of Real Grievances, A 117
vintage watches 184–5
von Brühl, Count 62, 63, 64
von Fersen, Hans Axel 101–2
von Sochocky, Sabin Arnold 156

Waltham Watch Company 127

warfare 137–9, 151–4, 162–4
watches/watchmaking
 in art and literature 90
 battery-powered 172–3
 child labour in 117–18
 digital watches 175–7
 Dutch forgery 76–8, 81, 85–8,
  89–90
 in Early Modern Europe
  41–52
 effect on other industries 65–6
 in eighteenth century 55–6,
  58–62, 65–6, 90–1, 115
 and industrialisation 126–9
 interwar innovations 159–60
 in London 55, 58–62
 luminous dials 154–8
 oldest known 32–5
 pilot watches 139–41
 quartz watches 173–5
 and relationship to time xiii–
  xiv, 91
 skull watches 36–41
 in Switzerland 87–8
 trench watches 152–4
 vintage watch revival 184–5
 in warfare 137–9, 151–4,
  162–4, 165–71
 women in 83–4

Watchmakers and Clockmakers of
 the World (Loomes) 77
water clocks 13, 14 see also
 hydromechanical clocks
Watson, Samuel 65
Weems, Philip Van Horn 140,
 167
Wellington, Duke of 109
Wilsdorf & Davis 148–9, 160
Wilsdorf, Hans 146–7, 148–50,
 160, 161, 163, 164
Wilter, John 76–8, 88, 91–2
Windmills, Joseph and Thomas
 88
women
 painting luminous dials 155,
  156, 157–8
 watches for 171
 in watchmaking 83–4
 working hours of 124–5
Wood, Henry 139
Woolf, Virginia 76
Worshipful Company of
 Clockmakers 48, 55–6, 126
Wyke, John 65

Yakubov, Zion 93
Yankee watch 127–8, 134, 179
Yeats, W.B. 188

About the Author

REBECCA STRUTHERS is a watchmaker and historian from Birmingham, England. She cofounded her workshop, Struthers Watchmakers, in 2012, with her husband, Craig. Together they use heritage equipment and traditional artisan techniques to restore antiquarian pieces and craft bespoke watches. In 2017 Rebecca became the first watchmaker in British history to earn a PhD in horology. Rebecca is one of the few remaining horologists in the world making timepieces from scratch. She lives in Staffordshire with Craig; her dog, Archie; cats Isla and Alabama; and Morrissey the mouse.